OVER THE LICKEY!

DONALD J. M. SMITH

DEREK HARRISON

PETER WATTS Publishing limited

ISBN 0 906025 71 0

Published 1990
by
Peter Watts Publishing Limited
Stag House, Gydynap Lane
Inchbrook, Woodchester
Gloucestershire, GL5 5EZ

Typeset and produced for the publisher by
Mopok Graphics, 128 Pikes Lane, Glossop, Derbyshire
Printed and bound in Great Britain

ILLUSTRATIONS

(Front cover) 'Garratt Over the Lickey'. 2-6-6-2 tank number 4999 (later British Railways 47999) storms over the Lickey summit with a northbound freight beneath a crackling December moon in 1937. These powerful locomotives were built by Messrs Beyer, Peacock in Manchester to the design of Sir Henry Fowler, the L.M.S. Chief Mechanical Engineer. This particular locomotive was latterly rebuilt with a revolving coal bunker to facilitate the fireman's task of feeding the massive fire grate. (From an original painting by George Heiron)

(Back cover, upper) A Cardiff-Newcastle express pounds up Lickey headed by three cylinder compound 4-4-0 No. 40925 and an unidentified BR Standard 4-6-0. What a lovely period shot this is. The old Deeley-designed compound adds atmosphere to this holiday-time train that would be crowded into the corridors with eager travellers, oblivious to the fact that they were making railway history. (T. E. Williams)

(Back cover, lower) A fine study of 'Big Bertha' descending the Lickey after yet another banking duty. The headlamp was for working the Incline in darkness to assist in judging distances when buffering up. The sighting bars affixed to the front running plate also helped when lining up a train's rear before the ascent. 21C was the shed code for Bromsgrove. (J. Edgington)

(Title page) Storming Through! The picture that says it all – a superb shot of 'Jubilee' class 4-6-0 45626 'Seychelles' pounding up the Lickey Incline with a Cardiff-Newcastle express. The date is unknown, but it may possibly be the harsh winter of 1962/3. 'Seychelles' was a regular locomotive seen on the Birmingham-Bristol route, along with several other members of the 'Jubilee' class. One hopes the photographer did not catch pneumonia! (J. Hyde)

Class 47/0 No. 47229 is captured by the camera climbing the Lickey Incline at Vigo on 8th August 1985, in charge of 6E55, a block rake of empty oil tank wagons returning to Port Clarence from the Unitank depot at Bromsgrove. *(C. J. Tuffs)*

Contents

Prelude

The Lickey Incline—mid nineteen fifties—one of my favourite railway locations. Whenever the Jubilee or Black Five hauled trains blasted their way up in the 1 in 37.7 towards Blackwell it was music all of its own.

Summer Saturdays in particular. A steady stream of holiday trains from West of England seaside resorts en route to the industrial cities and towns of the Midlands and the North never failed to bring dramatic effect to the Worcestershire countryside.

On a number of occasions the Lickey landed me in hot water—for being late home for Saturday tea! You tended to hang on for just one more holiday special. There was the chance it might have that seldom-seen express engine. Possibly a Baby Scot, the unrebuilt Patriot class 4-6-0s. The regular Patriot—you could almost set your watch by it—was 45509 'The Derbyshire Yeomanry'. What a pity that particular locomotive never survived. An important gap in steam traction heritage lies unfulfilled.

A memorable August Bank Holiday Saturday springs to mind. My parents decided, regarding late Saturday tea, that 'enough is enough'—long before Margaret Thatcher coined the phrase.

"If you can't beat 'em, join 'em," my mother added.

You can imagine my surprise when our 1937 Morris 8 saloon rolled into Blackwell Station yard, picnic tea and all! From 4.30pm until about 9pm we sat on the station platform seat devouring salmon sandwiches, jelly laced with orange slices, fruit cake (home-made of course), with a choice of tea or fizzy lemonade. My parents were fascinated by the trains they saw—the L.M.S. built Beyer-Garratt on a northbound freight impressed. The steady woof, woof, woof, woof from the Garratt's exhaust, followed by the staccato hammer-beat from 58100 'Big Bertha' banking the long train of wagons through the station. The weather was glorious—sadly yours truly did not take photographs in those days. You, the reader, can only share the memory in your mind's eye. The old Morris made numerous subsequent 'excursions' to the Lickey area—the magic had well and truly rubbed off.

On certain nights, if the wind was in the right direction, one could hear the trains being banked 'over the Lickey'. Dozing in bed, at my Northfield home, it lulled one into a secure slumber. There was a sense of peaceful serenity—Lickey would somehow become a guardian angel.

With the passing of steam the atmosphere went. The diesels could, and still do, climb the incline mostly unaided. Nowadays, the Inter City 125s roar up in fine style. A memory of the early 1980s was riding in the cab of a 125—we hit the bottom of the Lickey at 90mph to sail over the summit at 70! Times have indeed changed.

When Donald Smith approached me to assist him in putting the history of the Lickey together I was more than pleased to participate. The years somehow have simply rolled away—there I am once again at Blackwell or Vigo watching the trains.

Whilst researching the memories section the writer met up with Derek Barton, one time fireman on the 'bankers'. Would he kindly jot down any recollections? He did so. A week after receiving them his wife wrote to me—the final sentences literally rooted me ... "*On his way to a club meeting last Wednesday morning, he posted the envelope to you. He died two hours later.*" The ten pages he sent are indeed treasured!

If the reader finds nostalgia reigns hoof in mouth with history in the pages that follow, then so much the better. Those who remember the Lickey in its heyday will again relive the days of express trains, heavy freights, bankers, the brakesmen. Hopefully it will rekindle that spirit of many a loco crew who worked trains 'Over the Lickey'.

January 1990

Derek Harrison
Edgbaston,
Birmingham

Introduction

The town and manufacturing complex of Birmingham (a city from 1889) has long been the hub of an important transport system. From its centre radiate routes of railways, roads and canals, converging at the intersection of a St. Andrew's Cross. Yet, because Birmingham dominates a plateau of the central midlands, its approaches, especially from the south west, climb from the Severn Valley on an almost continuous gradient. This was the pathway of the Birmingham and Gloucester Railway, later to merge with the Midland Railway, of which it became an important main line, to be absorbed at the 1922/23 grouping within the newly-formed London, Midland and Scottish Railway. Since nationalisation, in 1948, it has been controlled by both the London Midland Region and the Western Region of British Railways, the latter's boundary presently extending as far north as a point between Blackwell, at the summit of the incline, and Barnt Green. A special signboard marks the spot.

The main obstacle between Gloucester/Worcester and Birmingham is a spur of the Lickey Hills, extending north to south west. This forms an outer escarpment, contiguous with the Clent Hills, as though guarding the Birmingham uplands and Black Country region. It may be noted that the Lickey Hills comprise a complex upfold of Pre-Cambrian and Cambrian rocks, rising through a later strata of Upper-Carboniferous rock and New Red Sandstone. It is a geologically important site revealing an underlying floor of younger sediments below this area of the West Midlands. It is across this spur of rising ground that the Lickey Incline or Bank, also known to railwaymen as the Bromsgrove Bank, has been constructed—mainly on embankments and through shallow cuttings. It extends between Bromsgrove Station (now a halt) at the southern end, and Blackwell at the northern end, for two miles and ten chains, falling towards Bromsgrove. The Incline is 1 in 37.7, being one of the stiffest and most punishing gradients on standard gauge trunk routes in Britain.

Until recent years and the dieselisation of the line, nearly all trains were banked from Bromsgrove, where special locomotives were reserved for these duties. This service is now infrequently used, with diesel-powered bankers assisting only certain classes of freight traffic, or occasional cases of mechanical failure. It has been, and is, the practice to drop the bankers at Blackwell, on the summit of the Incline, from which they return via the down line. The up line is considered as being towards Birmingham and Derby.

The greater part of the route, from Birmingham (New Street station), passes through pleasant suburbs, missing the more formidable aspects of local industry

Catch me if you can! Two ex-L.M.S. 'Jinty' tanks follow a southbound freight down the Lickey Incline during the late 1950s. Both of the engine crews seem happy to pose for the photographer! *(John Edgington)*

encountered on other urban approaches. For a considerable part of its journey it follows the southerly course of the Worcester and Birmingham Canal, not only in parallel but in close proximity, partly through tree-lined enclaves that bring a feeling of green space to densely populated areas. Railway and canal part company in the suburb of Bournville, although reunited beyond Bromsgrove near the foot of the Incline, which the waterway descends via 58 pound locks, 30 of which form the 'Tardebigge Flight'. passing through out-lying areas of Birmingham, especially King's Norton, Northfield and Longbridge (where there is a junction leading to the former Halesowen Branch), the main line skirts the huge complex of the British Leyland plant, originally known as the Austin Motor Works, which is the only evidence of large scale industry, apart from Cadbury's of Bournville (the factory in a garden), between central Birmingham and its south western boundaries.

Three and a half miles beyond Longbridge the line enters Barnt Green station, with its branch line to Alvechurch and Redditch. Then, after a further mile, Blackwell, where it makes an abrupt descent. Almost as abrupt is the change from town to country as, apart from a few scattered houses, the surrounding area, until reaching Droitwich or Worcester, is essentially rural. This is an unspoilt corner of the former county of Worcestershire, now merged into the new county of Hereford and Worcester. Where it is possible to see open country, through gaps in hedges and cuttings, there are north westerly hills skirting the Black Country, while to the south east an undulating landscape rolls towards the Northampton Uplands. On a clear day there is a wonderful panorama of the Malvern Hills, beyond Worcester, in a south westerly direction. Just after dropping from the summit, about three quarters of a mile to the east, may be glimpsed the expanse of Hopwood Lakes, two large reservoirs popular for weekend sailing.

Approaching the bottom of the Incline very little is seen of the town of Bromsgrove, apart from straggling suburbs, the town centre being at least a mile from the station by which it was served. At one time there was a large railway works and motive power department, partly at the rear of the down platform. This remained standing but derelict for a number of years, until demolished in 1981. Bromsgrove Station itself has undergone a great transformation within recent years, changing from one of the principal points on the line, with two platforms and a goods yard, to an unstaffed halt with single platform. Immediately beyond the up platform, in a south westerly direction, are sidings leading into a large oil storage depot, also sidings—on two levels—used by the Lickey bankers. Ordinary goods facilities have now been withdrawn from the area.

Modern High-Speed Trains and InterCity expresses take the Incline with an almost imperceptable change of pace. Travelling in darkness it is now fairly difficult to guess when the Lickey has been traversed—very different from the days of steam and the necessary, if brief, delays. In past times slow trains, unfitted freights and some excursions were frequently held back at Bromsgrove to let more important traffic take precedence. There was always a short halt either in or at the throat of the station, when heads would crane out of carriage windows, especially those of enthusiasts and people in the know, trying to spot the bankers.

The length of the Incline is measured from what was formerly the end of the Bromsgrove up platform, Bridge 97, which is 55 miles 7 chains from Derby. Worcester is around 200ft. above sea level, entailing an overall climb of some 700ft. to Blackwell Summit, the testing part being beyond Bridge 97.

An interesting view of the Incline, looking south west, was painted by a Victorian landscape artist, E. J. Dolby, of which numerous lithographic prints were made, dating from June 1840. This shows the almost carpetlike pattern of small fields, with mature trees and hedgerows, clinging to the lower slopes of the Lickey Hills. This has changed very little over a period of a hundred and fifty years. It is a scene typical of the English midlands, not spectacular in the grander sense but essentially rustic and peaceful, perhaps difficult to believe that, even at the time of painting, it was barely a few miles from the clanging 'Workshop of the World' and centre of a thousand trades.

January 1990

Donald J. Smith
Castle Bromwich
Birmingham

Chapter 1 — Early Days

THE BIRMINGHAM AND GLOUCESTER RAILWAY

As early as 1824, in the reign of George IVth, there had been plans to build a railway linking Bristol and Birmingham via Gloucester, also connecting with Tewkesbury and Worcester. This was to have been known as the Bristol, Northern and Western Railway, challenging the near-monopoly of water-borne services via the River Severn and tributary canals, that took at least six days between the main centres. The first meeting in this connection was held at the White Lion Hotel, Bristol, in 1825. It was backed, among other speculators and businessmen, by the Quaker brothers Charles and Joseph Sturge of Birmingham. Later the same year Josiah Jessop, son of the even more celebrated William Jessop (pioneer of the Surrey Iron Railway and the Grand Junction Canal), was appointed Engineer to the line. A professional surveyor, W. H. Townsend of Bristol, planned a route between Gloucester and Worcester but considered the northern section, between Worcester and Birmingham, so difficult—due to the need for deep cuttings or inclined planes (especially in crossing the Lickey Hills)—that work on this part was to be deferred indefinitely.

A start should have been made on a section between Bristol and Worcester in 1826, but this was a year of financial crisis, during which many subscribers wished for a return of their deposit money, with calls for abandonment. The untimely death of Josiah Jessop may also have been an important factor, regarded by many as an ill omen. Subscribers up to this period had been wary and moderate, reflecting a depression in the money market throughout 1826/27.

There is claimed to have been a revived interest for a line between Birmingham and Gloucester after the successful opening of the Liverpool and Manchester Railway in 1830, although further practical steps were not taken until 1832—in the reign of William IVth. The brothers Sturge were again prominent in arousing public interest. The plan was for '*making a railway from Birmingham to Gloucester, with a branch therefrom ...*'. The branch in question, intended as an important loop, was to serve the city of Worcester, diverging near Abbotswood, to the south of the city. The northern junction, although not then decided, was to be at Stoke Works, between Droitwich and Bromsgrove.

Surveys and suggestions for such a line were made by both Isambard Kingdom Brunel and George Stephenson. Brunel was then aged 26 and on the threshold of his highly successful career as a railway, civil and marine engineer. In 1833 he was commissioned by the brothers Sturge, by whom he was also assisted, to make a complete survey of the proposed route. This was his first railway contract. Observations were also made by George Stephenson, especially relating to the problem of inclined planes. Both Brunel and Stephenson agreed that a line at least three miles further east than the one eventually chosen would be necessary and appropriate.

These two surveys were considered in detail, Brunel's missing both Cheltenham and Worcester, to the great disappointment of many inhabitants of these centres, especially in Cheltenham. As first planned this was essentially a goods line, although the interest of passengers was—for a certain period afterwards—allowed to dominate. Another important reason for the proposed diversions was to avoid both a difficult bank and/or tunnel at the northern end of the line, the same problem facing Townsend nearly eight years earlier. Keeping engineering costs as low as possible by traversing flatter ground had long been an orthodox if unenterprising policy in planning both railways and canals. Yet by this time recession appears to have been halted—if only on a temporary basis—so that lack of funds was not the main consideration. In token of this, the area through which the line penetrated, also the centres it connected, were described as '*both prosperous and moderately progressive*'.

There was considerable discussion, and many advertisements, concerning the project from the summer of 1834, although an official prospectus was not published until 3rd October 1835. The scheme then became so popular, both nation-wide and in the West Midlands, that there was a danger of over-subscription, lists having to close much earlier than anticipated. The first share issue was £750,000 in £100 lots, which were disposed of, to the extent of 7,500 lots, almost before ink had dried on the paper. It is recorded that the last batch of shares, valued at £500, was sought by over 4,000 applicants. As with many early railway schemes, especially in the north and midlands, there was considerable backing from the Quaker community or Society of Friends.

The system was intended as a standard gauge line of 4ft.8½in. (then frequently known as narrow gauge). From the earliest days it claimed the moral support of several other early railway companies, hoping to check northward encroachments of the Great Western 'seven foot' broad gauge, when this finally appeared.

It was first decided, once a line across the Lickey Hills had been chosen, to construct two inclined planes between Birmingham and Bromsgrove. Cost of land needed for the construction of the whole line varied from £60 to £3,000 per acre, total purchase price anticipated as being in the region of £150,000. What seemed to be unrealistic demands were made by a small group of greedy landowners, but settled after a period of compromise and negotiation at Parliamentary level.

In 1836 Captain William Scarth Moorsom was appointed official engineer to the line, assisted by two former comrades of his army service. A new survey was made, entirely replacing Brunel's, and Captain Moorsom urged for the idea of a single incline at Bromsgrove, worked by locomotives throughout. This was in direct opposition to the findings of both Brunel and Stephenson. George Stephenson, in particular, claimed the mechanical impossibility of working steep gradients by adhesion methods, recommending the use of fixed engines and rope haulage for unavoidable planes, as used on Camden Bank near the approaches to the first Euston Station (Euston Square—terminus of the London and Birmingham Railway). Yet Captain Moorsom was a person of great individual character, determined to carry through his schemes without reference to public opinion or advice from even the highest authorities.

Captain Moorsom was a Yorkshireman by birth and descent, born near Whitby in 1804. Unlike most of his contemporaries in civil engineering he had neither come from a specialised engineering background nor risen from the ranks of labour. He had received a public school education and further studied at the Royal Military Academy, Sandhurst, where he took a special interest in surveying. Although commissioned in the 52nd Light Infantry Regiment he was essentially a surveyor and engineer, further renowned as an amateur landscape painter, with a good eye for country. After service in Nova Scotia and other parts of Canada, with subsequent visits to the United States, Captain Moorsom left the army on half pay, eventually living in a house owned by his father, near construction sites on the London

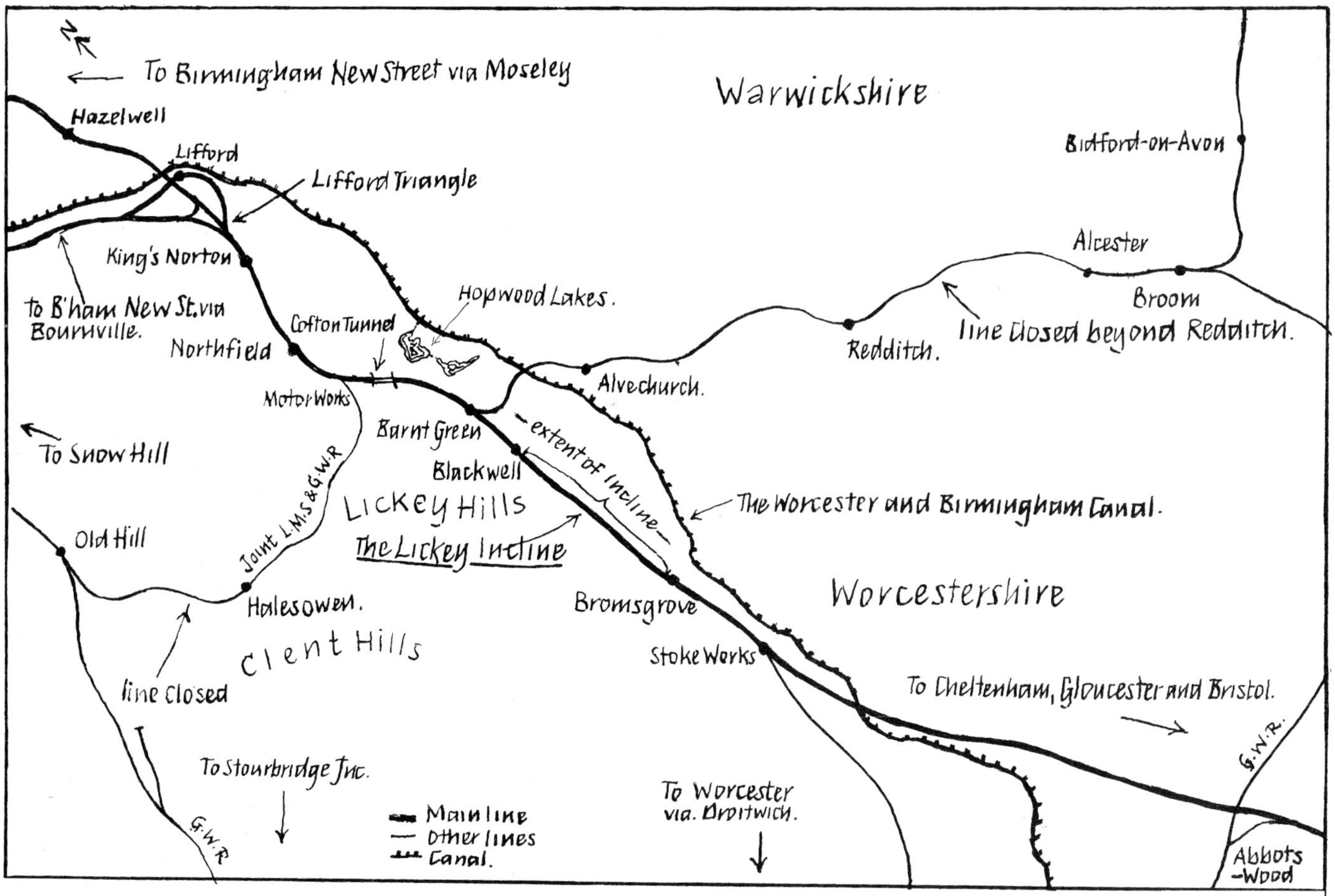

Sketch-map showing the position of the Lickey Incline on Birmingham-Gloucester line — not to scale. *(Donald Smith)*

and Birmingham Railway. Here he became both interested in railways and acquainted with the Stephensons—father and son—with whom he was later in disagreement concerning the Lickey Incline. It may be further noted that the brother of Captain Moorsom became not only an Admiral in the Royal Navy, but also Chairman of the Birmingham and Gloucester Railway, ending his career as Chairman of the London and North Western Railway, inheritors of the London and Birmingham Railway.

As Chief Engineer Captain W. S. Moorsom was paid a salary of £1,000 per annum. He eventually had four assistant engineers, each at a salary of £500 per annum, for setting out the land or course of the railway, there being two men each for the northern and southern divisions of the line. Such men were also responsible for the construction work in detail. Each assistant had a sub-assistant, at half their salary, able to make working drawings and report on general progress. The staff of the construction team further included a chief draughtsman and a clerk. A legal contract bound Captain Moorsom to work on only one other railway at a time, apart from the B. and G. Company.

Nearly all employees, except the engineering staff and clerks, were organised on para-military lines, common with most railway company workers up to that period. They wore uniforms similar to those of the police and obeyed a strict code of discipline, being frequently sworn-in as special constables. Foot patrols were organised along parts of the main line and inclined plane, during both construction and operational periods. Many of the officers of the company were ex-servicemen, frequently on half-pay and retaining their army or navy ranks as a token of authority.

Throughout the passage of the Bill for the Birmingham and Gloucester Railway there was constant harassment from conflicting factions in the Worcester area. The Grand Connection Railway, projected for a line between Wolverhampton and Worcester, felt confident enough to try excluding the B. and G. from Worcester territory, which caused great concern both locally and on the London Stock Exchange. Meetings were held at all levels, their outcome being an agreement that the B. and G. should only approach Worcester on G.C.R. terms or if the G.C.R. plans failed to mature—as eventually happened.

The first reading of the Bill before a Committee of the Commons was on 20th February 1836, its second reading being eight days later. The final reading before the Commons was on 6th April, before passing to the Lords. Yet even at this stage there was a protest meeting in Worcester, trying to force the B. and G. company to change its plans, not yet passed into law, but this failed through lack of financial backing. While one faction tried hard to bring the railway into Worcester, as near the centre as possible, the G.C.R. was concerned to keep it as far from the city as the law would allow.

The final route was planned via Cheltenham, Ashchurch, Spetchley, Dunhampstead, Bromsgrove and Moseley to the centre of Birmingham. Between Gloucester and Cheltenham use was to be made of a former horse tramway (later worked by steam locomotives transferred from an industrial system at Neath Abbey Iron Works), terminating in Gloucester Docks. Spetchley was considered the main stopping place for Worcester, three miles to the west of the line. (Upon opening a free service of horse-drawn coaches plied between Spetchley Station and the city centre, and was long regarded as a financial drain on the resources of the company).

There was only nominal opposition in the House of Lords, with favourable petitions made by Lord Ellenborough on behalf of the Mayor and Corporation of Gloucester, a petition on behalf of Birmingham Council by the Marquis of Salisbury, and a deputation from the Mayor and Borough of Tewkesbury. (It may be noted that a Tewkesbury faction, previously opposed to Bill, had been won over by the offer of a branch line connecting with the B. and G. main line at Ashchurch. Powers to construct this mixed gauge branch were granted on 15th April, 1837).

The Bill had been steered through the Commons by Lord Granville Somerset and through the Lords by the above mentioned Marquis of Salisbury. Having passed the Lords on 22nd April 1836, and receiving the Royal Assent two days later, the Birmingham and Gloucester Railway was duly incorporated by Act of Parliament. It was the first Railway Act of the session and the only railway, up to that time, to gain sanction at the first attempt.

The first meeting of the company, after the passing of the Act, was at a hotel in Birmingham on 19th September 1836. Alternate meetings were to be held in Birmingham and Gloucester with the first half-yearly meeting arranged for 1st February 1837. There were eighteen Directors and a Board of Management divided into two committees—one for each end of the line. Share capital was declared at £950,000, with rights to borrow £316,000 on the credit of the undertaking.

Working or engineering headquarters, during construction, were located in both Birmingham and Worcester. The main executive offices were in Waterloo Street, Birmingham. In Worcester the company occupied a large Georgian house in Foregate Street, near what much later became Foregate Street Station (Worcester and Hereford Railway), demolished when it was found necessary to build a new railway bridge taking a line to Malvern and Hereford.

Although the B. and G. promised Captain Moorsom a free hand in the layout of the Lickey Incline there was, for several months—even after the passage of the Act—a rumble of discontent among the shareholders. At the first half-yearly meeting a complaint from the floor, by a gentleman named Chance, seemed to express these doubts to a chorus of approval. It was decided to consult Joseph Locke, then Chief Engineer of the Grand Junction Railway, an acknowledged expert on such problems. After an extensive period of examination and study Locke reported in favour of Moorsom's plan, the facts—when published—greatly contributing to a swift rise in the value of B. and G. shares. It is almost certain, however, that if the opinions of either Locke or the Board of Directors had turned against him Captain Moorsom would have resigned on the spot.

Mr Samuel Baker, then Chairman of the B. and G. Company, in commenting on Lock's report, affirmed that *'if the route (chosen by Moorsom) had not been adopted, much traffic would have been lost which was now secured, and that fears with regard to inclines were visionary'.* Further reference was made, in the same speech, to the growing number of people wishing to trade and travel between Gloucester, Cheltenham, Birmingham and other centres of the West Midlands.

The first tenders for contract work were made in October 1837, concerning the supply of beechwood sleepers, to be treated for dry rot and preservation (kyanising) by the Anti-Dry Rot Company of Gloucester, to whom they were to be delivered on certain dates between February and May 1838. The double track permanent way, for the greater part of the route, was to be first laid as a baulk road, the flat-bottomed rails fixed to longitudinal beams or balks joining the treated transverse sleepers at intervals of several feet apart. In later years the whole system was relaid with transverse sleepers only. Ballasting was of several different kinds, including combinations of burnt clay, burnt marl, gravel, sandstone, rock marl and broken stone of both lias and oolite. Rails were of a shallow, parallel type with a broad bottom-web, bevelled at the sides, weighing 16lbs. per yard. There were also iron chairs on the longitudinal baulks, these latter being of American pinewood in 60ft. lengths. Over the Incline, however, and other gradients, baulk lengths were 15ft. with cross sleepers of beech or larch (mainly beech) placed much nearer in parallel than on the level.

Applications for brickwork, masonry, earthworks, fencing and ditching contracts were to be made directly to Captain Moorsom—so said advertising in Berrow's Worcester Journal in early 1838. Plans and specifications were on view at the offices of the Company in Worcester and Birmingham. While some claimed that Moorsom was taking a revolutionary step in dividing the work between various contractors for different projects and areas, this was merely reverting to the procedure of early canal engineers during the second half of the 18th century.

In January 1838 it was anticipated that the line would be opened north to Bromsgrove within eighteen months, including branches to Worcester and Tewkesbury, with three years needed to complete the contract, a forecast proving, in the outcome, unrealistic. Construction work was divided into fourteen separate contracts.

Within a short period most contracts were let, and sub-contracts arranged, for work at Cheltenham, Bredon, Eckington, Barnt Green, King's Heath and Moseley (then described as near Birmingham), confirmed by an article in Berrow's Worcester Journal for the second week in the New Year. It was claimed that the B. and G. would be one of the most important and profitable railways in the kingdom as Birmingham was then obtaining much of its supplies through the ports of Bristol and Gloucester. The line would also become an essential trunk route between the north and the west of England, especially as Bristol was also then the main trading outlet for the Americas. The B. and G. margin of profit was forecast, in the same article, to rival even the highly successful Grand Junction Railway, whilst the figure of £950,000 (£200,000 above the original estimate) was claimed to be sufficient for overall construction, although an even larger sum was eventually found necessary.

During the month of February Captain Moorsom was asked to provide at least three locomotives for use on the line, to enter service as from January 1839. Tenders were received from sixteen makers, on which reports were made to the Board of Directors during October 1837. Among the more promising was an offer from William Norris of Philadelphia, U.S.A., claiming that an 'assistant' or banking locomotive could be supplied for heavy gradients, especially useful on the Lickey Incline (the only gradient meeting this description on the entire route). A Norris 4-2-0 locomotive known as 'George Washington' had undergone special trials in hill climbing on its home ground, earlier the same year, of which Captain Moorsom may have been aware through newspaper reports and contacts in the United States. He was in correspondence with Norris from the late summer of 1838. Other locomotives, for general purposes, were eventually tendered for by established British companies, and it is certain that the first batch owned by B. and G. were of British origins.

The contract for earth moving and general construction work on the Lickey Incline was first advertised on 21st August 1838. Quoting the advertisement ... *'The total quantity to be excavated is about 240,000 cubic yards, to be carried to an adjoining embankment. The total length of the contract will be two and a half miles, and is calculated to be worked with self-acting planes. Drawings and specifications for the work, with printed forms of tender and conditions, will be ready for inspection at the offices of the Resident Engineer at Moseley, near Birmingham, on 3rd September 1838 ... endorsed 'Tender for earth works at Finstall' and no tender will be attended to, except in the printed form above referred to. BY ORDER. William Burgess, Superintendent and Secretary. Waterloo Street, Birmingham'.*

A young assistant engineer based in the Worcester offices at this time was Herbert Spencer, later to become one of the great philosophers of the Victorian era. A champion of the theory of evolution and author of 'Synthetic

Philosophy' in ten volumes, Spencer began his short career as a railway engineer in 1837 on the B. and G. system. He worked on both surveys and the construction of the Lickey Incline, often in close collaboration with Captain Moorsom, who he described in his diary as ..*'the best specimen of a gentlemen I have ever come near'*. In July 1840 he was employed in the reconstruction of a large under-bridge, taking a public road under the Incline in the direction of Bromsgrove, that did not at first meet with the requirements of the local authorities. This work had to be done in less than four weeks, after the pulling down of the old bridge, at a cost estimated between one and two thousand pounds. Spencer was later responsible for decisions made in the motive power department, although an outdoor life did not appear to suit him and he resigned from railway service, for health reasons, in 1846.

At a half-yearly meeting, held in February 1839, it was claimed that the greater part of the work was already under contract and the line making excellent progress. A section between Cheltenham and Bromsgrove was expected to open during the first half of 1840, if only for experimental runs. No fewer than 3,000 men were, at that time, employed on various work sites, with over a third on the Incline.

During the early summer of 1839 work began at the foot of the Incline, making a clearing for a station and enginehouse. A large workshop was also planned, this being on the eastern side of the line in the parish of Stoke Prior, at least a mile from the centre of Bromsgrove. Construction began in August, the station buildings having a booking hall, ladies' waiting room, general waiting room and offices for the Resident Engineer. This was to be described as a first class station or depot (there being first, second and third class categories on the line according to size and general facilities), on a single storey plan, with adjacent goods shed and several cottages for railway employees. According to Wishaw, the station was opened in 1840 as the centre of a small railway colony. It was situated in a 10ft. deep cutting, at the commencement of the Incline, also approached down a sloping drive from the main Bromsgrove Road, passing to the rear of the Station Master's dwelling and the gardens of several railway-owned cottages. A colonnade extended the whole length of the building, comprising four gabled porches. There were at first three tracks in the station area, the centre track reserved for banking engines—an additional fourth track was added when the Midland Railway gained control, keeping through traffic away from either platform. As with other stations on the line it was first known as a depot, in the American style, perhaps reflecting Captain Moorsom's earlier associations with North America. It was to be the main intermediate station on the route, designed in the neo-classical style, with most of the other stations—apart from Cheltenham—having buildings of Victorian-Gothic or neo-Tudor architecture. The platform, at the time of opening, was 10ft. wide at its narrowest point but only 18in. above rail level. Even taking into account that rolling stock was much lower than in modern times, such a low platform was a cause for constant embarrassment and complaint, although it was not raised until several years later, while under the control of the Midland Railway. B. and G. carriages, in common with those of the London and Greenwich Railway, were fairly low-slung, the wheels fitting into so-called 'paddle boxes' recessed beneath the floorboards.

In the official handbook of the Railway Clearing House, Bromsgrove was eventually described as a passenger and parcels station, also having a goods station (able to cope with furniture vans, carriages, portable engines or machines on wheels) and having facilities for horse boxes, prize cattle vans and horse-drawn carriages (pleasure or private carriages) delivered by passenger train. Crane power was five tons.

There were facilities for banking engines and a small locomotive shed at Blackwell, near the summit of the Incline, from 1842.

Locomotives

From the early days locomotives were divided between Bromsgrove and Cheltenham sheds. These included the Norris or American type, which were the first foreign locomotives to work for a British main line company. Captain Moorsom, in ordering them, may well have been influenced by reports of the performance of the engine 'George Washington' over a 1 in 14 gradient. He later claimed that he only negotiated across the Atlantic after two of the leading British firms refused to consider his requirements. Tenders from other British manufacturers did not specify anything powerful enough to cope with the Incline on a regular basis, although it has since been claimed that such work was well within the scope of home production. Moorsom, however, was under pressure from his directors to provide locomotives as soon as possible and therefore may not have been as discriminating as he should have been. In the meantime, four British designed and constructed locomotives by Forrester—numbered 1 to 4—for general duties, were delivered in 1839 at a cost of £1,600 each, these to be used on the line south of Bromsgrove.

The first locomotive ordered from the Norris works was No. 6, 'Victoria', a Class A Extra, at a cost of £1,750, minus tender. It could not be delivered as early as was first anticipated so William Gwynne, the Norris agent in Europe, persuaded B. and G. to invest in a

Norris-built 4-2-0 locomotive as supplied to the Birmingham and Gloucester Railway circa 1840.

(Donald Smith)

smaller Class B engine costing £1,545, complete with tender. It was to be known as No. 5 'England', arriving in March 1839 and tested on the Grand Junction Railway. Results were not as encouraging as they might have been, although purchase was confirmed. However, six improved Class B engines, Nos. 7-12, were subsequently ordered, as were three more heavier Class A Extra engines, Nos. 13-15.

The first Class A to be delivered, 'Victoria', was tested on the Bolton and Leigh Railway, but results were again unsatisfactory, due to the state of the track. About this time five more Class B engines appear to have been ordered through Nasmyth, Gaskell and Company (Nos. 26-30) and four from B. Hick and Son (Nos. 18/19/24/25).

The Class A Extra locomotive No. 13 'Philadelphia' came in May 1840. She was the only engine then shedded at Bromsgrove able to cope with the Lickey Incline on a regular basis. She arrived in the care of a specially trained driver and engineer named John Donahue, by whom she had been accompanied across the Atlantic. Typical of the Norris types she had 4ft. driving wheels and 2ft.6in. bearing or carrying wheels, being of 4-2-0 wheel arrangement. The bearing wheels were under the smokebox while the driving wheels were nearer the firebox, the former having a 3ft. wheelbase. All wheels were of cast iron, as were most other parts of the machine. The firebox was of the domelike or haystack type, as used on the Midland Counties and several other railways of the period. Inclined outside cylinders were 12in. diameter by 18in. stroke. Weight was approximately 12 tons. Characteristic features included swivelling bogies at the fore-end, on a truck frame, for easing round tight curves. Fuel burnt was coke, carried in a small four-wheeled tender with outside coke rails. Water on the tender was stored in a horseshoe tank surrounding the bunker space. Like other Norris types she was fitted with iron tubes and firebox which soon had to be replaced by copper and brass at considerable expense. There was also highly effective sanding gear. History tells us that Norris engines proved to be very costly in terms of running and regular upkeep.

In the first test on a gradient of 1 in 37.7 'Philadelphia' hauled a load of 75 tons at a speed of 10mph for just under the measured mile. In later trials with a Bury locomotive of the London and Birmingham Railway, with Mr Bury on the footplate of the rival machine, she proved greatly superior, especially in coping with heavier loads.

Rolling Stock

Early passenger rolling stock of the B. and G. Company consisted of First, Second and Third class carriages, First and Second class vehicles were provided—in 1839—by Messrs F. and C. Arthur of Birmingham and William Shackleford of Oxford. First class carriages were 21ft.6in. over the buffers, normally seating eighteen passengers in two full compartments and two half or coupé compartments, also having seats for brakesmen, or guards (one faced in each direction of travel) at imperial or roof level. Second class carriages seated 24 passengers in three full-sized compartments, also having roof seats for brakesmen and open-topped luggage racks, the latter not at first appearing on First class vehicles. These Second class carriages, with a slightly shorter wheelbase, were 15ft.6in. in length, while the width of all coaching stock was 7ft. All compartments in Second class were open to each other above the level of certain partitions. Windows were not glazed in Second class until October 1844. Buffers and draw gear were operated on a spiral-spring principle, with double brake shoes between each pair of wheels.

For several years Third class vehicles ran open to the sky, without seating, being little better than goods stock. They had so-called standing berths or stalls in double rows at each end, with a luggage compartment in the centre. Passengers facing in the direction of travel, also Second class passengers in unglazed compartments, were advised to wear goggles or spectacles, frequently of fine wire mesh, hired or sold for the purpose. By 1843 there were covered Third class carriages, in which people might either stand or sit. Some Third class stock, with proper seats, was also reserved for women and children.

Other items of coaching stock, running with most passenger trains, were horse boxes, open carriage trucks and luggage trucks. Roof racks were later introduced on First class carriages and the luggage trucks converted to Second and Third class accommodation. All stock was four-wheeled.

The Royal Mail was frequently carried in roof racks that were covered over (sheeted up) and protected from the elements. Special mail vans were introduced at a later period.

The livery of coaching stock was dark buff lined and picked-out in black. The arms of the company appeared on the

First Class coupé carriage, Birmingham and Gloucester Railway circa 1840. *(Donald Smith)*

centre doors of First class carriages only. Trains on the B. and G. were the first to have communication between footplate and guard by means of tubes and whistles.

All goods stock in the early days was of a four-wheeled type with dumb buffers and simple 'hook-on' couplings. Wagons for general merchandise had five plank sides, weighing 2½ tons tare and being 10ft.3in. long over the headstocks. At the onset there were 60 ballast wagons with low sides, each weighing about two tons and costing £65 ex-works. Many of these latter were used in constructing the Incline and on general work sites. It is claimed that special coal and mineral wagons of the modern type, designed for the purpose, were first used for mainline traffic on B. and G. metals. Carriages and other road vehicles were conveyed on open, flat trucks, while there was also a wide variety of both closed and open wagons for livestock.

Opening the System

A party of directors, leading shareholders and their guests made an informal inspection trip over the line, between Cheltenham and Bromsgrove, on 1st June 1840. Their journey began at 9am, reaching Bromsgrove at 10.41am, after a brief stop at Eckington and a fifteen minute halt at Sketchley. They were greeted on the platform of their destination by a brass band hired by a Mr Bayliss, then in charge of work on the Incline. There were so many flags and buntings that local people were misled into thinking it was the grand opening ceremony, which did not occur until some three weeks later.

On Monday 15th June 1840, an important sale was held in a field at Stoke Prior, next to the main line. This was conducted by a local auctioneer, Mr W. Higgs, and mainly consisted of surplus materials, tools, barrows, etc., used in the construction of the Incline, Bromsgrove Station and other contracts. Large numbers of sleepers and deal planks were on offer, subsequently being used by local farmers for fencing material.

The line between Cheltenham and Bromsgrove, of 31½ miles, was opened to the public on Wednesday 24th June 1840, special arrangements being made for Worcester and Birmingham passengers to transfer to road coaches at Spetchley and Bromsgrove respectively, at the expense of the B. and G. Company. The first public train, drawn by one of the Norris engines, had two First class and two Second class carriages. This was again an affair of bands and flags, music provided by the band of the local Volunteer Rifle Corps. The train left Cheltenham ten minutes late, arriving in Bromsgrove at 10.50am and, after a ten minute run-round, returned south by 12.27pm.

In normal service there were to be four passenger services each weekday (but, in the early days, no Sunday services) running as follows:-
Bromsgrove to Cheltenham 10.30am and 7pm.
Cheltenham to Bromsgrove 9.00am and 6pm.
Three extra trains were soon added in each direction. First, Second and Third class fares were 8/-, 5/6d. and 8/6d. respectively.

Passengers were not allowed to smoke, either on the trains or in the station buildings, as there were strict rules concerning this (and spitting) on company property. Railway staff were expected to be polite and helpful at all times but not to expect tips for services rendered. If a train was full to capacity longer distance passengers, after a ticket inspection, were given priority over those merely booked to the next two or three stations.

It was calculated at this time that when a Camp Hill terminus was opened at the Birmingham end, a through train would take about two hours, with a further fifteen minutes to Gloucester, the standard gauge permanent way between Cheltenham and Gloucester then being under construction. Several correspondents of newspapers, magazines and journals, especially representing the Birmingham Gazette, wrote to praise the comfort and safety of the line, although not perhaps travelling Third class. One journalist, with experience of frequent travel on the London and Birmingham Railway, the Great Western Railway and the London and Greenwich Railway, claimed that coaching stock on the B. and G. was far better than anything he had encountered up to that time, also that arrangements '*for the opening of the line were admirable ... and there was no confusion anywhere, nor did any accident of the slightest description occur*.' Not all, however, were in such an enthusiastic mood as a contributor to Herepath's Railway Journal, signing himself 'Commercial', felt the suspension was far from perfect and that leg straps should be introduced to prevent lighter people from being shaken on to the floor.

Opening the Incline

The inclined plane, later known as the Lickey Incline, was opened to through trains on 17th September 1840. There was a station at Blackwell near the summit, opening on the same date, also a temporary terminus known as Cofton Farm—usually shortened to Cofton. This latter was near what later became known as Cofton Tunnel also known as Groveley Hill Tunnel, eight miles from the centre of Birmingham, to which it was linked by a service of horse-drawn coaches, operated until a larger terminus at Camp Hill, in the Birmingham suburban area, could be built. Cofton Tunnel was the only tunnel of note on the line, being some 440 yards long, cut through New Red Sandstone and lined with engineering bricks throughout. (It was opened-out as a cutting, with concrete-lined embankments, over the weekend of January 26th-28th 1929).

Regarding safety measures for working the Incline, a house was provided near the summit where a trustworthy person could reside, to oversee that safety regulations were obeyed to the last detail. While on the trains this person would be in full control, having authority above engine driver, guard or any other servant/official of the company. In those early days there seems to have been considerable anxiety among members of the public, which the presence of such an official did much to allay.

An early working over the Incline was described as follows:-
'*it commences near Bromsgrove Station, where an additional engine is attached. The average speed up the bank was between 14 and 15 miles per hour. The journey went smoothly and without delay...*'. On a return trip it was noted that ...'*the brakesman (as he is called) took his station on the middle carriage, the engineer and guard being under his control and regulating the brakes etc, in compliance with signals from him*'.

The average train did not descent the Incline at more than 16 miles per hour. It was understood that no passenger train would travel down at more than 20 miles per hour, but much less in slippery or foggy weather. There was eventually lineside accommodation for two pilots or senior brakesmen, one acting in a reserve capacity, which implied that up and down trains alternated and never passed on the bank. According to a slightly later report under brakesmen, also known as travelling porters or bank riders, were assigned to ride on each train, there being one on each passenger vehicle, under the direct supervision of the head brakesman. Goods and cattle train were limited to a maximum speed of 10 miles per hour, either up or down the Incline. The first trains to meet and pass on a regular basis were night mails on a service dating from 6th February 1841. Full details concerning bank workings at this period are included in a later appendix.

The Bromsgrove Messenger for 19th July 1925 carried an interesting article on the Incline, illustrated with a reproduction of Dolby's famous painting, previously mentioned. This showed 'Philadelphia', one of the first and most powerful of the Norris engines, taking eight loaded wagons up the gradient at a speed of 9¾ miles per hour, total weight being 74 tons. In the same column reference was also made to the memoirs of Mr Joseph Guise, a

Lickey Incline circa 1840: a freight train of the period climbs towards Blackwell behind a Norris built 2-4-0. An interesting painting this as the train, consisting of eight loaded trucks, would prove too much for a locomotive of this size to haul without the aid of a banking engine. What are the people up to walking freely on the Incline one wonders! *(Collection British Rail)*

Bromsgrove centenarian, who claimed to have made a Third class journey over the Incline in 1840. At that period there were no seats in Third class, people having to lean or squat against the sides of an open vehicle. Tickets were, at first, hand-written on thin paper and many travelling in cheaper parts of the train never intended paying—if they could help it—swarming over the platform and into the empty vehicles as soon as they appeared. Mr Guise recalled that there were numerous free trips up and down the Incline on the day of opening, as part of the celebrations, which may have been a publicity boost to prove the safety of rail travel, especially 'over the bank'.

After the opening of the Bromsgrove Railway Works there was a general shortage of accommodation in the town. Mr Guise and many others taking in lodgers—on a temporary basis—to ease the situation. At Finstall Park, on the lower part of the Incline, the arch of a large railway bridge was partly bricked-up at the ends to provide makeshift dwellings.

Signalling—of the manual type—was in the care of railway policemen, nicknamed 'Bobbies', the apparatus they controlled having double vertical columns in parallel, of a unique design, with oil lamps and reflectors.

Trains were seldom, if ever, banked from the rear under B. and G. control, at least not until the improvement and standardisation of buffing gear. The only exceptions appear to have been in time of emergency when the train or pilot engines failed. The majority were taken to the top by the special engines designed for the purpose. Eventually most trains were double-headed, a pilot engine being attached in front of the train engine, between Bromsgrove and Blackwell. On nearing the summit the fireman of the pilot climbed to the rear of the tender (a vehicle much smaller than its modern counterpart) and released the couplings, the train engine having shut-off steam. The pilot then shot ahead into a nearby siding guarded by hand operated points, the lever of which would be thrown back before the arrival of the assisted train, with only moments to spare. It is amazing that there were no serious accidents in this context, especially as it is certain that neither train crews nor pointsmen were paid danger money. In later years a trip hook or 'connecting catch', invented by J. E. McConnell, was installed on the footplates of all pilot engines. This was operated by means of a foot pedal, saving the fireman perilous encounters with his coal/coke supply and the laws of gravity!

There is an amusing account, dating from the early days of operation, that records how a special train of May 1841 was mishandled on the Incline. This was a trip organised by the Mechanic's Institute of Gloucester and District, including a fair proportion of members from Cheltenham and Tewkesbury. By the time the train left Ashchurch—the nearest picking-up place for Tewkesbury—there were 263 passengers in 23 coaches, hauled by two engines double-headed. The train divided at Bromsgrove, the first part of fourteen coaches headed by the train engine and a banker. Although appearing to make a good start they stalled half way up the Incline and had to be assisted from the rear by the second portion.

Until 1841 a part of the line at Blackwell, used when dividing or checking trains, was on a slight gradient of about 1 in 300. Although tested and considered safe in normal conditions, some vehicles with their bearings well run-in could be moved by a heavy gust of wind or accidental prod, resulting in an escape down the Incline, especially if not fully braked/chocked. In March there was a near miss with a run-away wagon that careered into Bromsgrove Station. This incident lead to a lengthening of the dead level section, near the summit, to 240ft. before coming on to the actual Incline.

Accidents and Calamities

Considering the number of serious accidents on many early railway systems the B. and G. were perhaps more fortunate than others. Passengers were usually well cared for at all stages and between 1840 and 1846 only one death was recorded in about two million journeys. However, Railway employees and some officials were more vulnerable than passengers because of the nature of their work, although frequently running unnecessary risks through flouting the rules. The extreme youth of some employees, especially in the early days of the railway, may account for a few deaths and injuries. Among these were three boys employed in construction work on the Incline in 1839. One was aged thirteen and the others fourteen, all of whom fell to their deaths beneath the wheels of wagons being gravity shunted to a spoil tip. One of the older boys was described as earning fourteen shillings per week, then a good wage by comparative standards, but the sole support of his invalid parents.

The first passenger casualty on the B. and G. was not on the actual railway but concerned a road coach owned by the company, used as the connection between Cofton Farm and the centre of Birmingham. A young woman passenger was standing near the coach, which she was about to enter, when a sudden or unexpected noise caused the horses to bolt. She was knocked down and fell under the churning wheels, only then to die within 48 hours. At about the same period a gentleman's groom was killed in Bromsgrove Station, having assisted in loading two of his master's horses into a box, to be afterwards attached to a train on the up platform. The box had to be manhandled on to the main line and was struck a glancing blow by an engine travelling in the opposite direction. Unprepared and unseeing the groom was knocked off balance and fell beneath the vehicle he was attempting to guide. He died of his injuries three days later.

Many accidents were predicted on the Incline, long before the system was opened to traffic, but few of these occurred. Some were caused by runaways, as previously mentioned, although even these were not as frequent as might have been expected. According to a column in the Gloucester Journal, during February 1841, railwaymen in Bromsgrove Station were amazed by the sight of a single wagon laden with quarry stone, rushing down the Incline towards them at a furious rate. It passed them at an estimated speed of 100mph, while they were still undecided as to their course of action. The wagon was later found intact three miles beyond the station area near Stoke Works.

During the summer of 1841 a rake of several wagons, left unattended at the summit, began to roll down the Incline but were brought under control by a nine year old boy who scrambled aboard the leading vehicle and attempted to apply the brake. His calls for help soon attracted his mother who also climbed aboard and, between them, they brought the runaways to a dead halt.

At this period wagons were sometimes run down the Incline, in controlled circumstances, with a person to check the brakes and shout warnings. This, however, was strictly forbidden after an elderly woman was killed on a dark evening, by a wagon without lights or warning signals of any kind. The woman in question had been crossing the line as a short cut to her own cottage, after domestic service at a house on the other side of the tracks.

The Explosion in Bromsgrove

About two months after the opening of the Incline there was a serious accident in the area of Bromsgrove Station, although sometimes recorded as having happened on the Incline. This involved an experimental engine working over B. and G. metals but not owned by the Company. It was an 0-2-2T locomotive (perhaps one of the first tank engines in Britain), with several unique features, and was designed by a Doctor Church of Birmingham, an American gentleman, who was also concerned with the design of self-propelled road coaches. He was considered something of an eccentric genius. This engine, first known as 'Eclipse', had been constructed by Inshaw of Birmingham for a private owner or lessee, Mr S. H. Goddard, and had already worked on the London and Birmingham Railway, under sub-contract, as a ballast engine. It made frequent appearances at Vauxhall shed and station in Birmingham, although usually unattended and uncovered on a dead-end siding. Driving wheels were 6ft.2½in. and trailing wheels 3ft.; there were also double frames, outside cylinders and inside bearings. In March 1838 she was offered for sale in the open market, but could not find an interested purchaser.

Two years later, her name changed to 'Surprise', she was offered to the B. & G. for possible work on the Incline. Therefore, from 3rd November 1840, there was an arrangement between Goddard and B. and G. for 'Surprise' to run over the bank 'with the trains', for a period of 28 days. During the afternoon of 10th November the locomotive had moved into the depot yard on a siding directly behind the down platform, having completed her daily stint. She was then under the control of Driver Henshaw and his son Paul, the younger man acting as fireman. After a few minutes they were approached by Joseph Rutherford, the Works Foreman, in company with Thomas Scaife—an off-duty driver—both of whom came across to inspect the machine at close quarters. Henshaw invited the newcomers on to the footplate and, because space was restricted, both he and his son dismounted, making extra room. As soon as Scaife and Rutherford clambered aboard, at exactly 5.30pm, there was an almighty uproar, the boiler bursting in clouds of scalding steam and a scatter of fragments.

Scaife was killed instantly and his body projected 60 yards before rebounding 25 yards—as though 'shot from a cannon'. The mangled corpse was unrecognisable as anything human and would have been hard to identify in different circumstances. Rutherford was also thrown some distance although partly checked by three brass rails at the side of the footplate that broke under the force of impact. Although less seriously mutilated than Scaife, he suffered multiple injuries and died next day, without sign of recovery. Of the onlookers Paul Henshaw was badly burnt and scalded but recovered after hospital treatment. Several of his workmates, passing through the yard, were hurled into the air and across the tracks, suffering cuts, burns and bruises of a more-or-less serious nature. The funnel of the locomotive bounced over a 100 yards while the iron top of the firebox embedded itself in a grass bank, well beyond the station area. A loose spanner from the footplate soared like a rocket, falling through the station roof, ceiling and floorboards to bury itself deep in the foundations.

The cause of the explosion was said to have been the thinness of the boiler plates and possible lack of boiler stays. Giving evidence at the inquest Joshua Horton and Brierly Hill, engaged to assemble the boiler, claimed to have examined the plates after the accident but testified '*they were not the type or thickness he originally ordered for the locomotive*'.

Scaife and Rutherford were buried side-by-side in the graveyard of Bromsgrove Parish Church. Both headstones appear to have been carved with representations of Norris engines in profile, which have been the cause of much innocent controversy over the years. 'Surprise' was very different to the machines depicted on the graves and, in explanation, the public have been led to think that the mason or sculptor involved—perhaps unacquainted with locomotives—had copied the wrong type, either from illustrations or artefacts. It is more likely that the headstones were of standard design and American manufacture—or produced under licence from an American firm. These could be obtained through mail order, chosen from a catalogue, especially favoured when erecting memorials to those killed in railway accidents or involved with railways during their working lives. The stones would be engraved by a mechanical process, leaving blank spaces on the lower part for a suitable epitaph.

The inscription on the gravestones in St. John's churchyard, Bromsgrove, are as follows:-

Tombstones at Bromsgrove Churchyard, commemorating the death of Thomas Scaife and Joseph Rutherford who lost their lives as a result of a boiler explosion on the Incline. This engine was not one of the Norris engines which was carved on the grave stone. *(British Rail)*

The tombstones after a period of neglect. *(S. Widdowson)*

Further Openings and Developments

A through line between Cheltenham and Gloucester was opened on 4th November 1840. There appears to have been very little ceremony but large crowds thronged the platforms to witness the first departure. Typical autumn weather prevailed with cold gusts and heavy rain.

On 17th December the line opened to Camp Hill, after which Cofton Farm automatically closed and intermediate stations opened at Longbridge, Lifford and Moseley. The final cost of construction was mentioned as being upwards of £1,266,000, far higher than the original estimate and a colossal sum for the early Victorian era.

Locomotives in service at this time were Nos. 1-21. Number 17, 'Pivot', remarkable for its odd name, came new from Norris through the agency of Joseph Banks and Company earlier in the year. All were employed on general duties, apart from No. 14 'Boston', No. 13 'Philadelphia' and No. 21 'William Gwynne'—the latter three working as bankers only.

By 1841 the Bromsgrove Works had greatly expanded. Many wagons and a few locomotives were known to have been built on the premises but greater efforts were directed towards repairs and maintenance, exclusively for the B. and G. The internal layout of the plant and locomotive depot, although at first based on a central turntable with six radiating lines or outlets, was entered by means of a single track only, until later improvements were made by the Midland Railway. (It was under the M.R. that Bromsgrove became the main wagon repair centre of the entire 'Midland' system.) The locomotive running shed, at first forming part of the same building, was the principal motive power centre on the line. In later years, however, the only locomotives shedded at Bromsgrove were bankers, apart from a single engine kept for local trip work and emergencies. A brass foundry, a blacksmith's shop, a turnery, a carpenter's shop, a paint shop and a stationary engine house were all found under the same roof or adjacent.

The first Superintendent of the Works and M.P.D. was William Creuze, assisted after the untimely death of Rutherford by a man known as Walworth, who became his personal friend and drinking companion. It was claimed, at least in hindsight, that important repairs for which these men share responsibility, were frequently neglected as a result of their convivial habits. This careless attitude led to a further serious accident (described later), as a result of which Creuze died of his injuries.

By February 1841 C. R. Moorsom emphasised an ever-increasing need for smartness, discipline and economy, in true naval fashion. There were vigorous attempts to attract a wider variety of traffic, especially freight, which, despite earlier aims and intentions, the B. and G. had neglected in favour of passengers. The years 1841/42 were, unfortunately, a time of slump and passenger receipts began to fall, although goods and minerals greatly increased to off-set and increase in running costs. From October 1841 large quantities of coal and coke were sent to a new depot at Alstone Lane, Cheltenham, a line of business that continued to flourish and expand until the end of the century.

From the mid-1840s large numbers of livestock and cattle trains were run over the Incline, mostly on Tuesdays, Thursdays and Saturdays. There were several important markets held in the West Midlands on these days, with a great demand for farm animals, store cattle and butcher's beasts. Livestock trains were usually headed by two, three and even four engines, each train consisting of about 50 closed or semi-open trucks. Sheep wagons had rows of vertical bars on both sides, with two decks or levels, which could be loaded independently over special ramps. Drovers frequently travelled in the vehicles with cattle and pigs, preventing them from slipping down or helping them to rise, when fallen. Beasts that fell during the journey might otherwise have been trampled to death causing general panic and needless suffering.

Under the Chairmanship of C. R. Moorsom the Board of Directors was reduced from twelve to five members, they being Messrs Daniels, Sturge, Gibbins, Tymbs and himself. Only Sturge had survived from those attending the first Board Meeting in 1836. Under this new regime each of his Board members represented an executive department—'motive power', 'permanent way', 'coaching and carrying' and 'finance'.

At the first half-yearly meeting of this era it was revealed that a planned extension to link with the London and Birmingham Railway was well under way, its completion promised for either May or June of the same year—1841. (A reciprocal exchange of goods traffic with the Grand Junction and Birmingham and Derby lines was not arranged until 1844). It was also noted at the above meeting that over 120,000 ticket holders had been carried over the line, without the slightest accident or injury to a single travelling passenger.

On 23rd March a meeting held in Birmingham discussed the possibility of Sunday local trains, in addition to the needs of the General Post Office. There were representatives from many different bodies, both churchmen and laity, most of them deploring not only the need to make profits on 'a day of worship and tranquility' but to break the Sabbath and deprive workmen of their lawful rest. It was eventually decided to work one Sunday train in the morning and one in the afternoon, between Birmingham and Gloucester, throughout the year.

Plans to make a junction with the Bristol and Gloucester Railway had been abandoned at an earlier period, as there had been lack of general agreement concerning the branch to Worcester from Abbot's Wood, over which it was said the B. and G. seemed to be 'dragging its feet'. It was claimed that the Birmingham and Gloucester were slow to fulfil promises concerning this branch, over which the Bristol and Gloucester anticipated certain privileges and running rights. Talks were, however, resumed at a later date and amalgamation became a meaningful reality, leading to even greater mergers.

The Second Explosion.

Despite the new Chairman's emphasis on greater discipline and efficiency, affairs at the Bromsgrove Works had sadly deteriorated, reaching their nadir during the month of April 1841. According to a report in the local *Bromsgrove Messenger:*

'On Wednesday morning (April 7th) an accident caused the death of one and serious injury of two or three others, by the blowing out of a plug from the boiler of one of the bank engines. William Creuze, the engineer of locos' was dreadfully scalded and died about twenty-four hours later. Two or three persons who were on the engine and tender at the same time, although injured, were recovering. The fireman of the locomotive, Edward Shaw Carter, was burned on hands and face.

'He (Carter) said he was employed on the Tuesday night on the 'Boston'. He left the station (Bromsgrove) at 10.50pm to bank one of the trains up the incline, accompanied by Creuze and Israel York (the driver). On their return York stopped the engine at Vigo Bridge and went home leaving Carter in charge of the engine and asking him to wait for Creuze who had also got off. He remained parked on the bridge for about two hours until a watchman informed him that the next train was due in five minutes. He then ran the engine down to Bromsgrove Station where he crossed over to the other line and got the tender filled with water. He afterwards took up Walworth the foreman and his wife and second brakesman, and ran the engine back to Vigo, where the Walworths and the second brakesman got off, walking along the Incline. He then ran the engine to the top of the Incline to refill the boiler and ran her back again. He stayed about twenty minutes, after which Creuze, Walworth and his wife and a man named Archibald Torrey,

Foreman of Messrs Nasmyth and Company, came and got on the engine and tender. He took off the brakes and they started down the incline at between 7 and 8mph, the balance indicating a pressure of 46lbs per square inch. He knew that this was the exact indication for he had altered it himself from a higher pressure after they had ascended the incline the first time. The steam was just beginning to blow when they started, whereupon Walworth lifted up the balance and allowed the steam to escape until there was not enough power to raise the valve. When they had gone about three hundred yards, the water and steam began to emit in great quantities and with great violence, through the firebox on to Creuze and the others and caused the firebox door to spring open. Carter, after a couple of minutes, jumped off the loco' which ran another short distance then stopped. He then went back towards the engine meeting Walworth and his wife and Torrey on the way. They found Creuze on top of the tender, but on being spoken to, he made no answer. Walworth took off the brake and they ran the loco' back to Bromsgrove Station where Carter went for help. The engine had only recently been repaired but had not suffered any accident before. In the repairs that had been made, a copper stay had been introduced through the boiler in place of one of the tubes but this, Carter thought, would give extra security.

'Thomas Horton, an engine fitter, said he had taken the engine to pieces for the purpose of sending it to the boiler makers and on return, he had put it together again. A stay was introduced through the boiler from the tube plate in the firebox to the tube plate in the smokebox. The stay was there for the security of the tubes and to keep the plates from being drawn into the fire. He had made the repair under the direction of Walworth. The copper stay was fastened by means of iron nuts screwed at each end, the stay being 1⅛in. diameter. It was the same thickness and was fastened in the same way as others he had put into boilers. The engine had been fitted with a new firebox and had started work on the Thursday before the accident. On the following Wednesday evening Carter had examined the boiler of 'Boston' and found that a short tube that had been put into the boiler plate of the firebox had been blown out. It had been put there in place of one of the tubes which had been taken away in order that a hole was left in the smokebox and to blow mud out of the boiler. It was fastened by means of a long steel ferrule and a man named Morgan had helped him to plug the hole. It was only fastened by being driven in with great force and it would have to resist the pressure of the steam by the force with which it had been driven in. Creuze died at 5am the following morning. He had been educated at Cambridge and was respected and liked by other officers and servants of the company, it was stated. A verdict of accidental death was returned and a deodand of £25 placed on the engine, which incidentally was repaired and put back into service.'

Memorial tablet to William Creuze B.A. — first Superintendent at Bromsgrove Works and Chief Engineer of Locomotives, who died due to another boiler explosion on 8th April 1841.
(Bromsgrove Museum)

Further Changes.

After the second serious explosion, two new men were appointed to run the locomotive and repair works at Bromsgrove, these being James Edward McConnell, Locomotive Superintendent—noted for his locomotive designs—and G. D. Bishop (trained as mechanical draughtsman), the latter acting as running foreman and general assistant. Both safety precautions and the need for economies were given greater emphasis than hitherto. According to rules laid down by McConnell, banking engines had to be in steam for at least sixteen hours at a stretch, but only eight trips per engine could be made over the incline during this period. Great savings were achieved, among other things, by altering the firebars and firebox frames, especially on the Norris types, which had been designed for wood rather than coke burning.

J. E. McConnell, a character in his own right, was never an enthusiast for the Norris engines, claiming they were far too light for their duties. 'Philadelphia', 'Boston', 'William Gwynne', 'Niagara', and 'New York' were all converted to tank engines late 1842. Even Captain W. S. Moorsom, their main champion, is said to have become disenchanted over the years.

From 17th August 1841 the Birmingham terminus at Camp Hill (named after the site of a Royalist military camp during the first English Civil War) was relegated to the status of a goods depot, although a through-surburban station of that name was retained until recent years. Most passenger traffic was then worked into Curzon Street, the terminus of the London and Birmingham Railway. Access to B. and G. lines from Curzon Street was over an extension between Camp Hill and Gloucester Junction. Eight trains per day were soon running in each direction but, at that period, only mail trains ran on Sundays. (The first Royal Mails were carried over the Incline into Birmingham on 14th January 1841 whilst Mail connections with Cheltenham-by-rail came on 5th February 1841).

During 1842, as from 1st May, trains up the Incline were handled by bank engines only, the train engine coming off at Bromsgrove, returning south after a short period on shed. A further locomotive would be waiting at the summit level to continue the journey into Birmingham, while the bankers went on shed in a specially designed building at Blackwell. At this time the yard at Blackwell was greatly improved with gas lighting and better refuelling facilities. Blackwell and Bromsgrove Stations were also connected by the telegraph.

'Great Britain'.

An important locomotive to appear in 1845 was 'Great Britain', designed by McConnell about twelve months earlier. It was an 0-6-0ST with outside cylinders of 18in. x 16in. and coupled wheels of 3ft.9in. diameter. At the time of building it was the most powerful engine in the British Isles. One of its more unusual features was an oval-shaped boiler barrel, the vertical diameter 3ft.10in. and the horizontal diameter an inch less. Total weight, in full working order, was 30 tons. It has been claimed as one of the few engines constructed from start-to-finish in Bromsgrove Works.

'Great Britain' entered service in June 1845. Its first number was 38 in the B. and G. stock list, although acquiring the number 276—then changing to 105—when acquired by the Midland Railway, who then later rebuilt it, under the supervision of Kirtley, as a well tank. Cylinders were also moved inside and reduced to 16in. x 24in.. It was again renumbered in 1855 to 221, but withdrawn in 1861. Although reported as broken up in 1862 it is thought that many parts were used in the construction of a new engine that remained active on the Incline until 1901. At one stage of its early career, it was recorded as having taken a train of 150 tons up the Incline at an average speed of 10mph, unaided.

Facilities for engine building at Bromsgrove Works were never as satisfactory as might have been expected. Only one or two were thought to have been completed, starting from scratch. Bury and Company, Nasmyth and Gaskell, Sharp, Robert Stephenson and others all supplied engines ready built, while after an amalgamation with the Midland Railway Company there were contributions from the well-known firm of Stothert and Slaughter. A useful engine of 1843 was the 0-4-2 No. 35 'Wadborough', from Sharp Brothers, which rendered reliable service until the end of the century.

Amalgamations

A through service from Birmingham to Bristol commenced in July 1844. A special opening train of 22nd July started from the Birmingham end on the stroke of 9am. Passengers continued their journey to Bristol—with break of gauge at Gloucester—by crossing to an adjacent platform in Gloucester Station. The first printed railway tickets were used on this route during 1845, about a year after the opening.

As from 14th January 1845 the Birmingham and Gloucester and the Bristol and Gloucester Railways merged to become the Birmingham and Bristol Railway, although separate Boards of Management were retained, termed either the 'Birmingham' or the 'Bristol' Committees respectively. Large numbers of new locomotives and items of rolling stock were introduced, both custom-built and second hand. After twenty months of operation both boards were absorbed into the much larger Midland Railway Company, the actual date of amalgamation being the 3rd August 1846. From this period the journey from Birmingham to Gloucester and Bristol often commenced in the Midland Station of Lawley Street, the latter eventually (like Camp Hill) to become one of the large goods depots in the West Midlands. Most of these services were transferred to Curzon Street from 1st May 1851 on a temporary basis, as a much larger joint-station, to be shared with the London and North Western Railway, was planned to open at New Street, nearer Birmingham town centre. New Street, however, was not opened until 1854.

During the brief early history of the Bristol and Gloucester line it had been dominated by the 7ft. or broad gauge system of Brunel and the Great Western Railway. A timely amalgamation with the Birmingham and Gloucester Railway and, through this, with the Midland Railway, saved it from the clutches of the Great Western and further developments of the broad gauge in a northerly direction. In the meanwhile the Birmingham and Gloucester had been under lease to the M.R. from 1st July 1845 to 3rd August 1846, a rent being paid to the M.R. of six per cent per annum of the united capital.

The break of gauge at Gloucester was highly inconvenient, especially for through goods traffic but, while changes were authorised south of Gloucester from 1848, there were very few through trains before 1854. The main broad gauge tracks in the Gloucester area, partly operated as a mixed gauge system, especially between Standish Junction and Gloucester, were not lifted until 1872, while a 7ft. gauge line remained to Parkfield Colliery, north of Bristol, until 1882, over which coal trains of the Bristol and Exeter Railway worked on a daily basis. By 1855 there were seven weekday stopping trains each way between Birmingham and Bristol and three on Sundays, all having to work over the Lickey Incline.

During the early 1840s Bromsgrove Station was connected to the town centre by means of a horse-drawn omnibus service, in addition to the use of cabs and semi-open carriages known as station fliers. The prototype omnibus seated ten inside with seven (and some luggage) on top. It had been commissioned by the management of the Golden Cross Hotel, although running costs were shared with the railway company. The makers were Marston and Brown, celebrated coach and carriage builders of Birmingham. It met all trains but was originally reserved for 1st and 2nd class ticket holders only. At a slightly later period other passengers were charged 4d. for an each-way trip. From 1st November 1856 horse buses of the Midland Railway also made connections between Redditch and Barnt Green via Alvechurch, Barnt Green being not too far from the summit of the Incline. A branch line to Redditch did eventually open in 1859, at first as an independent line, known as the Redditch Railway, then absorbed into the Midland Railway in 1863. Its junction with the Birmingham and Bristol line was at Barnt Green.

On several days throughout the spring of 1870 there were tests on the Lickey Incline concerning a patent braking system, then being considered by the Midland Railway Company. This was to be known as the Clark Chain Brake. It aroused great interest in Bromsgrove and the surrounding district, drawing crowds of onlookers to witness the tests. The brake, however, did not prove an unqualified success and was abolished, under terms of the 'Railway Regulations Act', before the end of the century.

From May 1871 a new line was opened from the outskirts of Birmingham at King's Norton, on the Birmingham-Bristol route, to a terminus at Granville Street, where many trains formerly using Curzon Street were redirected. This was along a system known as the Birmingham West Surburban Railway over which the Midland Railway had running rights. The joint-station at New Street was not widely used by the Midland Railway until a special extension to it (almost a station in itself) had been opened in 1885, linked to the Birmingham West Surburban line by a series of tunnels. The B.W.S.R. was absorbed by the M.R. in 1875.

On 18th April 1874 a special train of Pullman stock traversed the Lickey Incline and entered Bromsgrove Station. The chief passenger, an honoured guest of the Midland Railway Company, was George M. Pullman, inventor of the luxurious coaching stock named after him and a distinguished American citizen. He was accompanied by several high-ranking officials of the railway company and paid an extensive visit to the carriage and wagon works at the foot of the Incline. Vehicles in the train were of a type widely used by the M.R. at a slightly later period, having end-vestibules and clerestory roofs in the American style. Mr Pullman may have considered the suitability of the Bromsgrove Works for the repair and manufacture of his rolling stock, although the exact purpose of his visit was not recorded.

Despite the dangers of banking over the Lickey Incline there were, as previously noted, remarkably few accidents or mishaps, especially in Midland railway days. One of them occured at Blackwell, near the summit, on 2nd March 1877, when a heavy northbound freight train rammed the rear portion of another similar train. The first train contained numerous barrels of beer and ale, most of which was destroyed in the collision by running to waste. It was later claimed that the driver of the rearward train engine did not see the signal set against him. Normal services were delayed for several hours, on both tracks, many wagons being either shattered or derailed, with considerable debris on the up line.

Banking engines, throughout this period in the late nineteenth century, had to be turned at regular intervals to prevent uneven wear and tear. This was done either on the Lifford Triangle, near King's Norton, or at Worcester, between Foregate Street and Shrub Hill Stations. Both alternatives meant expensive and time-wasting trips, with the engines concerned out of service. A second-hand turntable was eventually acquired from Hendon (on the London St. Pancras-Bedford line), and installed at Bromsgrove in 1892. This had been 42ft. on the original site but was extended by 4ft. at its new location. It was not, however, suitable for the larger bankers that came after the First World War.

A shed list for 1892 shows five 0-6-0STs, designed by Johnson, at Bromsgrove, this wheel arrangement being the mainstay of banking duties under M.R. domination. They remained from the 1880s until replaced by a later series during the early 1900s. Both tank and tender engines were used, either direct from other parts of the system or with slight modifications. Several tank engines were converted into well tanks, which had also been the fate of 'Great Britain' the B. and G.'s last locomotive.

During the early years of the 20th century trains were getting both longer and much heavier, even on a railway such as the M.R., noted for its smaller trains and lighter engines. Under-powered locomotives and their stock frequently had to be assisted up the Incline by as many as four small tank engines. This was far from economical or satisfactory and several designs were drawn up for a larger banking engine. The majority of bankers were then Johnson tank engines, some nearing the ends of their working lives. One solitary example appears to have been of Kitson origins.

About the year 1910 there were plans at Derby for a 'super banker' with an 0-10-0 wheel arrangement. This was to have been a side tank locomotive with a fuel bunker capacity of 3½ tons and a water capacity of 1,500 gallons. The grate area would have been 25 square feet. Other designs that did not mature included an 0-6-6-0 tank engine at least 45ft. in length, and another articulated machine expected to tip the scales at 122 tons. Most of these plans and ideas were confined to the drawing office at Derby, nothing more concrete being considered in this department until 1912.

Special delivery! A Midland Railway express banked by a Kirtley outside framed goods engine, along with a Midland 0-6-0 tank, pounds towards Blackwell. The vehicles at the rear of the train are horseboxes – one of the grooms is leaning out of the window, no doubt curious to observe the photographer. The date is 27th July 1911.
(National Railway Museum)

Early in that year there was a special committee, chosen from the Board of the M.R., to discuss the design and construction of a large tender engine for banking, later to be known (unofficially) as 'Big Emma' or 'Big Bertha'. Plans, although shelved for two years, were meant to be kept strictly secret until (shortly before the outbreak of the First World War), the order for this machine was made official on 3rd May 1914, although its production was postponed by the war.

* * * * * * * * * * * * * * * * * *

A Tragic Accident

At two minutes to five on the morning of 5th October 1912, there was a fatal accident on the footway crossing in Bromsgrove Station. A porter, named Oliver Hawkins, was struck down by a returning banker off the Incline while taking parcels to the down platform. Engine crews had been warned to look out for other railwaymen crossing the line at this point, but it had been an exceptionally dark morning and, for some reason, none of the lamps had been lit on the down side of the station. A fellow porter saw or sensed the approaching banker and shouted a warning but his voice was drowned by the noise of other trains, while clouds of billowing steam further obscured vision. At the inquest a verdict of accidental death was recorded (Hawkins died instantly) with suggestions from the jury that a special over-bridge or subway should be installed for luggage and parcels exchange between platforms.... but this, although considered, was not acted upon.

These two designs for banking locomotives were considered by the Midland Railway prior to the First World War. (Derek Harrison Collection)

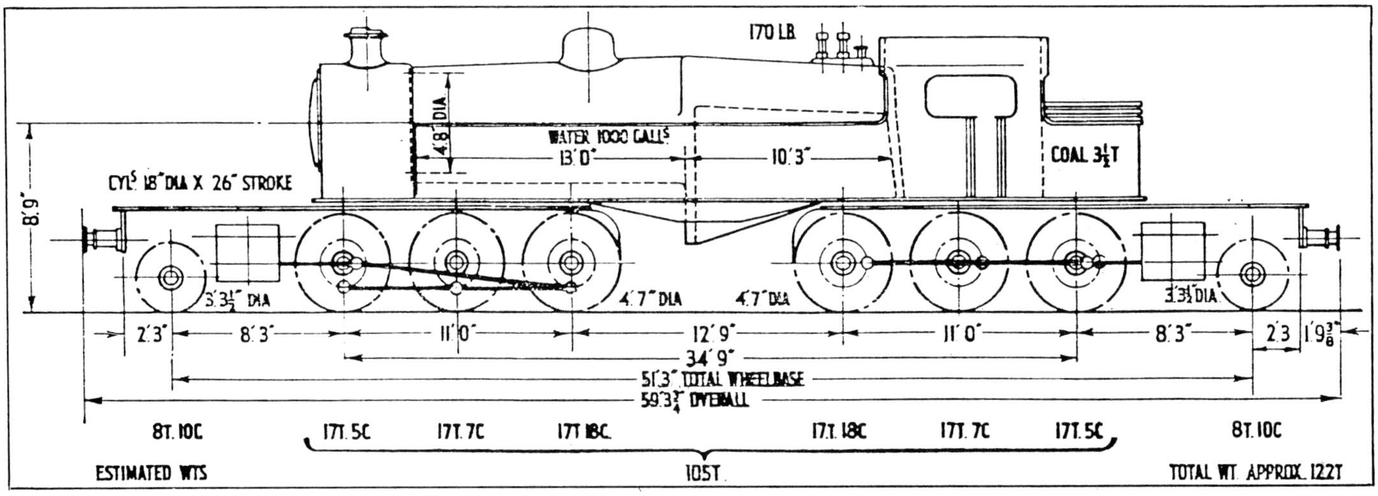

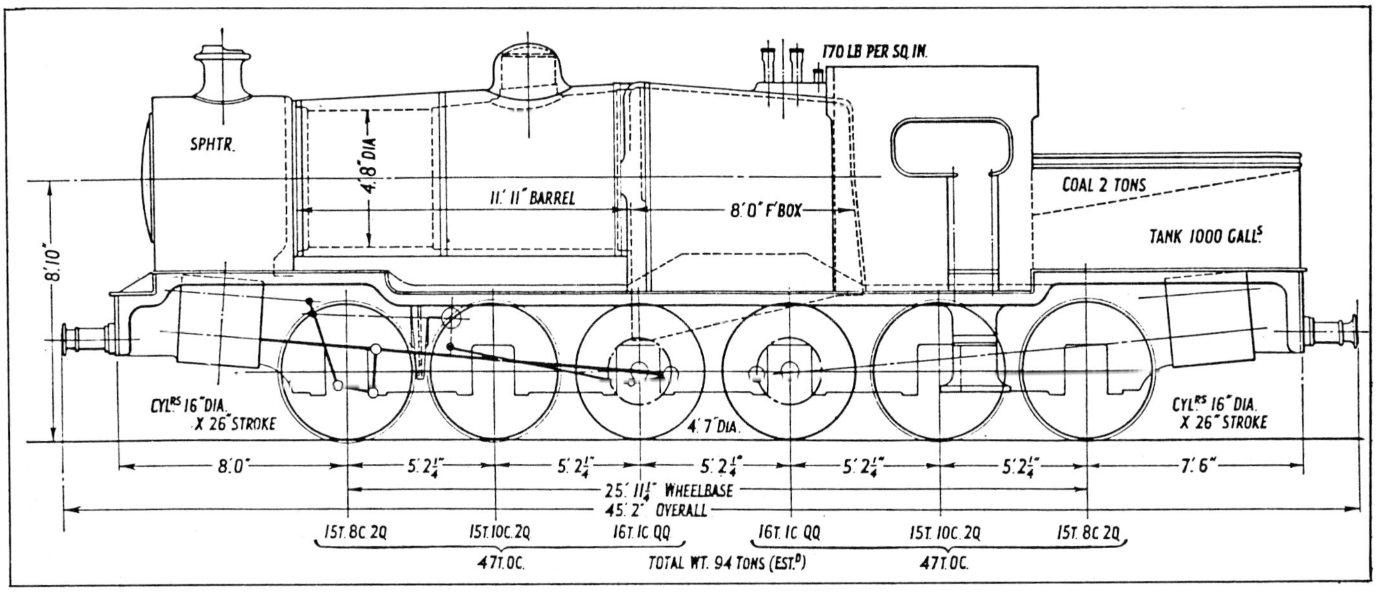

Two's company. A Johnson single-wheeler pilots 6ft. 9in. 2-4-0 number 197 on an up express on 27th July 1911. A fine selection of David Bain clerestory coaches form the train, including a twelve-wheeled dining saloon. The banking engine is not identified – almost any engine from Bromsgrove shed was made available for banking duties before 2290 arrived in 1919.
(National Railway Museum)

Blackwell summit on the same day. A northbound heavily-laden express roars over the top of Lickey in fine style, hauled by Midland Railway class 3P 4-4-0 number 779 double headed with a Matthew Kirtley (latterly rebuilt by Samuel Johnson) class 1P 2-4-0. Note how neat everything is, the tidy ballasting on the permanent way, cleaned glass in the gaslamp tops, the garden complete with bench seat, all laid out in apple pie order. The circle on the signal arm helps to date the picture. This type of semaphore marking was common on British railways before the familiar stripe, white on red, black on white, became universal. The embankments were also kept trimmed to prevent fires caused by sparks from locomotive chimneys.
(National Railway Museum)

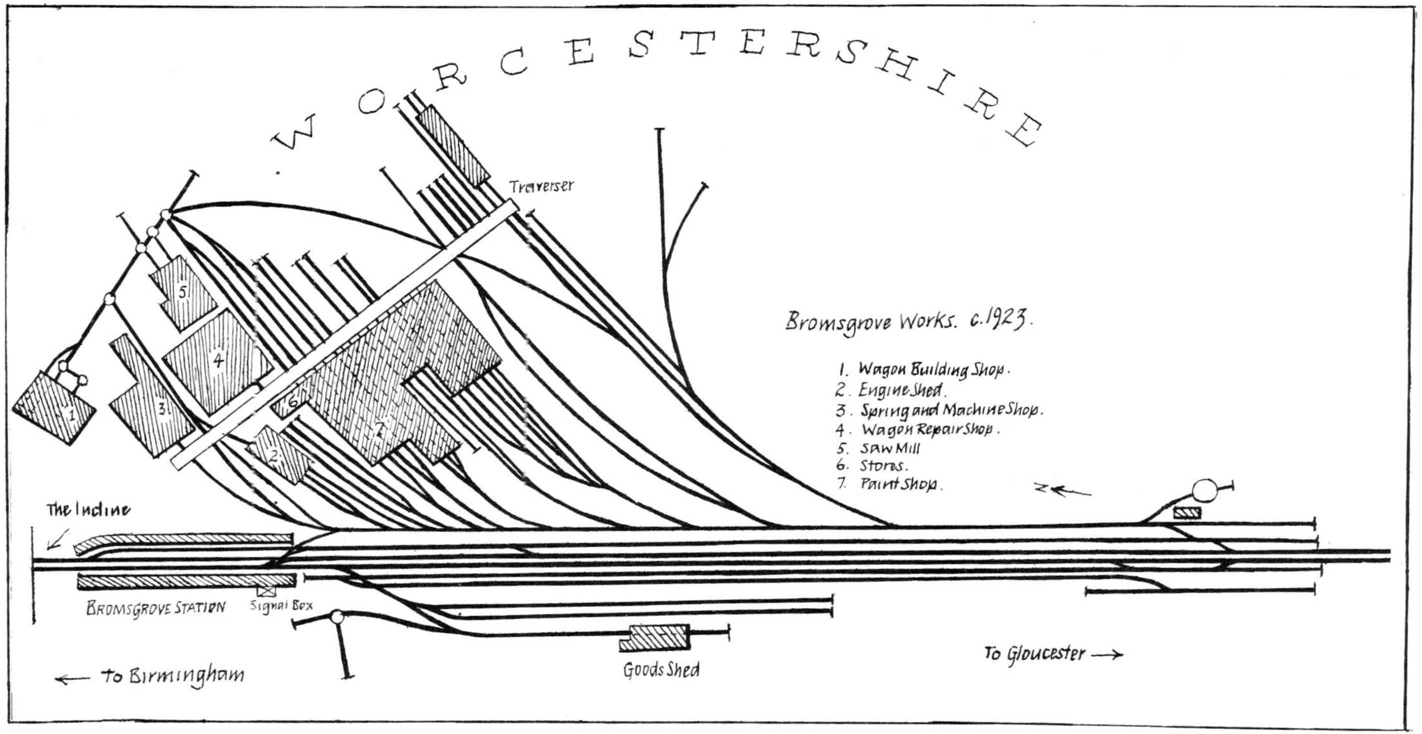

The track layout at Bromsgrove circa 1923 including the wagon works, now demolished. (Donald Smith)

Track layout of Blackwell Station (Lickey Summit), Midland Railway circa 1910. (Donald Smith)

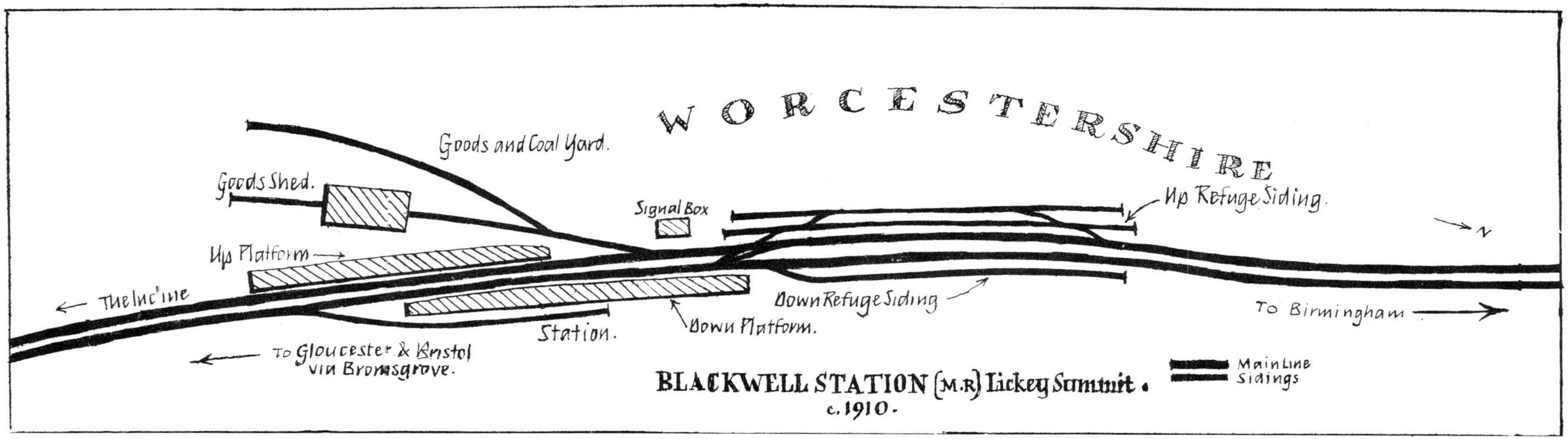

Chapter 2—1914 to the Second World War

Under construction! 0-10-0 number 2290 nearing completion at Derby works in 1919. The wheels are jacked clear of the rails to allow the motion to be fitted along with valve setting. The picture certainly gives the impression of a sizeable machine. The boiler needed to be capable of producing a tractive effort of 40,765lbs. in order to cope with the heaviest of trains working over Lickey. *(National Railway Museum)*

Between the late summer of 1914 and the autumn of 1918 large quantities of war materials were moved over the Lickey Incline, along with troop trains, hospital trains and specials with civilian refugees from battle-stricken areas in Belgium and France. These were serious challenges for the time-worn Johnson tank engines and further emphasised the need for something more powerful in residence at Bromsgrove, especially as many ordinary trains were then approaching 250 tons. But although there was a pressing need for new motive power and 'Big Emma' was already on the drawing board this, like many other projects, had to be kept in the background for nearly five years. Even the railway workshops were engaged in the supply and repair of gun carriages, tanks and military transport. Anything not directly connected with the war effort was thought unworthy of serious consideration.

Work commenced on the super banker shortly after the war ended. Many of those who had served in France now re-christened her 'Big Bertha', after the massive German railway gun that shelled Paris during the conflict. Officially she was un-named and known to her makers as No. 2290. 'Big Bertha' was first brought to public notice in January 1920, although unveiled at the Derby Works several weeks earlier in December 1919. She was a decapod, or 0-10-0, and only the second of this wheel arrangement to run in Britain. Important distinguishing features were an enclosed or covered tender for reverse workings down the Incline in all weathers, a massive electric headlamp (powered by its own generator) necessary for buffering-up in dark or foggy conditions, and inclined cylinders set at an angle of 1 in 7. She was one of the few engines on the Midland Railway to make use of Walschaert's valve gear, the only other exceptions being a batch of dock tanks introduced by Deeley in 1907 for use in the Bristol area. A tender engine was chosen rather than a tank engine, to prevent the extra weight of fuel and water carried being too great a burden on axles and main frames. Although some advocated a larger tender it was pointed out that 2290 would rarely move from the Incline, at the base of which coal and water were always ready to hand in large quantities. Owing to the steepness of the Incline, all wheels were braked. At the time of building the decapod was the largest locomotive ever used on the Midland Railway and the most powerful in the United Kingdom. The engineer and designer was Henry (later Sir Henry) Fowler, Chief Mechanical Engineer of the M.R.

During the 1920s most of the Johnson 0-6-0Ts of the 1900 class remained at Bromsgrove for banking, in addition to 'Big Bertha', taking duties turn-and-turn-about. Most acquired Belpaire fireboxes during the early 1920s, also Ramsbottom safety valves.

During the autumn of 1920 there were several rain storms, leading to landslips on or near the Incline, especially at Blackwell. Scores of tons of soil and clay fell across the tracks so that most drains were blocked and flooded. Hours of delay were caused by the need for clearance and reconstruction. Only a few years earlier the lines had been blocked by large trees brought down in spring gales.

In 1921 there was a serious coal shortage, due to a strike of miners at national level. Number 2290 was converted to oil-burning, although changed back to solid fuel a year later. It may be noted, in this context, that there had always been a preference at Bromsgrove Shed for firing with Yorkshire coal.

2290 at Bromsgrove in ex-works condition during 1921 after a temporary conversion to oil-burning. The headlamp had recently been fitted. Note Midland Railway signal gantry. *(J. Stirrup Collection)*

In January 1922 the first proposals were made concerning the main line groupings, which were finalised a year later. This led to the amalgamation of the Midland Railway with the London and North Western Railway, the Caledonian Railway and several other smaller groups. The new company (for many years the largest joint stock company in the world) was known as the London, Midland and Scottish Railway.

The last official excursions—a feature of railway travel introduced and developed by the M.R.—were run over the Incline on behalf of the Midland

A scene the authors would have loved to have witnessed! Newly delivered 0-10-0 'Big Bertha' number 2290, banks a Bristol-York express on 30th April 1920. The magnificent rake of David Bain clerestory coaches is hauled by an unidentified Midland Railway 3P 4-4-0 which is almost at the end of its climb up from Bromsgrove. The sanatorium at Blackwell dominates the skyline. Two shire horses in the foreground have obviously seen and heard it all before! The grass is too tasty to bother with trains. *(National Railway Museum)*

Almost there! A heavy northbound freight nears the summit of the Lickey Incline on the same day. An elderly Johnson 0-6-0 tank – its Salter safety valves lifting in the spring sunshine – gives support to a recently built 'Jinty' tank as the heavily-laden goods train struggles towards Blackwell. Quite a number of the open trucks are roped and sheeted, a specialised craft not seen on our railways in the 1990s. *(National Railway Museum)*

Railway in late December 1922. The last day excursion from Bromsgrove to Birmingham (New Street), at a return fare of 2s. 3d., ran on 26th December, posters and handbills in this connection signed, for the last time, by the retiring General Manager of the M.R., Frank Tatlow.

With the grouping there were numerous changes at all levels, although these were more inclined to favour the former M.R. operating and administrative staffs than those of their main rivals (but enforced allies) of the L.N.W.R. — once the self-styled 'Premier Line'. This led to an even wider variety of both freight and passenger locomotives to be seen on the Incline, although ex-Midland types always predominated.

At 1pm on Monday 14th December 1926 a goods train from Gloucester to Birmingham passed through Bromsgrove on the up line. It was a long and heavy train needing two bankers at the rear. As the forepart was seen to begin its ascent of the Incline a Mr Purdy, the manager of a station book

A fine study of Blackwell Station in 1922 showing original layout with Midland Railway nameboard.
(Real Photographs/D. Sharpe Collection)

Kirtley 0-6-0 number 2616 blows excess steam through the Salter safety valves approaching Blackwell in July 1922, the penultimate year before the grouping of British railways into four main line companies. The eight open wagons contain coal from South Wales.
(Real Photographs/D. Sharpe Collection)

Midland Railway 2-4-0 number 164 storms towards Blackwell on 29th July 1922. The first and fourth vehicles are Clayton non-corridor coaches. Trains in those days were made up of rolling stock from various sources within the relevant railway company. *(H. C. Casserley)*

stall, was in conversation with an off-duty driver. Attracted by a strange bumping noise Mr Purdy noticed that the tail-end of a wagon was sticking in the air. Both men ran the full length of the platform in pursuit, shouting their discovery to the crew of the last banker. Steam was shut-off and brakes applied in quick response—after a warning blast on the whistle—the train at last grinding to a halt. Three wagons had been derailed but were still fairly upright, these being an empty mineral wagon, a wagon laden with stone or ballast and an empty petrol tanker. The permanent way had been ploughed up for a distance of 50 yards but fortunately the derailed stock did not foul the down main line as an express passenger train soon made its appearance, approaching from the opposite direction. Thanks to the quick thinking of Mr Purdy and his companion a far more serious accident was averted. A breakdown gang came out from Saltley (Birmingham) and all workings were diverted to the down line until 6.30 the same evening.

A period shot taken in 1924. Midland Railway 0-6-0 number 3066 nears Blackwell summit with a heavy freight train. The second wagon contains pit props en route to the Cannock coal fields. Although taken in early L.M.S days, the locomotive buffer beam is still lettered MR. *(Real/Photographs D. Sharpe Collection)*

Midland Railway-built 2-4-0 number 96 pilots 2P 4-4-0 number 525 on a West of England flier in 1925. *(Real Photographs/D. Sharpe Collection)*

Torquay-Leeds holiday special express hauled by a hard-working 2P 4-4-0 number 514 with a motley collection of rolling stock. The first two coaches are former Lancashire and Yorkshire, followed by David Bain clerestories, all wooden bodied, of course, in those days! The year is 1925. *(Real Photographs/D. Sharpe Collection)*

In April 1927 several trips were made over the Lickey Incline by an experimental steam-turbine driven locomotive, constructed at the Gorton Works (Manchester) of Beyer-Peacock to the designs of L. Ljungstrom. This was a 2,000hp machine with a total length of 60ft. and overall weight of 140 tons, capable of 80mph top speed. Wheels were small and uniform like those of later diesel locomotives, without distinction between driving and bearing wheels. The coal bunker was concealed within the bodywork of the locomotive, at its rear end, containing 16 tons of fuel. Economical on coal, especially at high speeds, it was a near silent runner without the usual smoke and puff. Although the engine passed its tests with flying colours it was strange that it was never adopted by a main line company for regular services. On 14th April it successfully hauled a twelve coach service train between Derby and Bristol.

On various occasions in 1929/30 an ex-L.N.W.R. 0-8-4T was tested on the Incline, for banking duties, but was not considered suitable. This was engine No. 7953.

In 1933 about 150 trains were passing over the Lickey Incline each day. This did not include specials, empty stock movements and light engine workings. During the early 'thirties there was a change from ordinary semaphore to colour-light signals. Track was also quadrupled and greatly extended at both ends, making for improved services and swifter running times between centres. As part of this dramatic change a large signal gantry at the south end of Bromsgrove Station, spanning all tracks, had its signals carried above rather than below the main structure, for better sighting. Also with increased and heavier traffic the Incline had become a bottleneck, so a colour-light signal was placed half way up its length. Two trains could then proceed in an up direction at the same time, as though in procession. If the second train was travelling faster than the first then it would be checked at the midway signal, as need arose.

In a newspaper description of a then recently constructed signal box at Blackwell, opened in April 1930, it was termed *'one of the most technical show places on the L.M.S.'* While in those days most signalmen faced the running lines behind a row of levers, the operative at Blackwell turned his back on busy traffic, noting its movement on a train describer that occupied the opposite wall...*'on this were small red, yellow and green lamps labelled 'train offered' and 'train accepted', while others indicated whether a particular express, slow passenger train or freight is 'leading' along the line or 'following' another. The describer is worked in conjunction with a similar device in Bromsgrove Box at the bottom-end of the Incline. Also in the Blackwell Box is a large electric diagram which shows the progressing position of trains from the moment they enter the Incline to the moment they pass out of it.*

'When the Bromsgrove signalman offers Blackwell colleagues a train, he signals by bell and by means of the describer. The whole of the signalling system is semi-automatic. It is inter-

Horsebox special. 2P 4-4-0 number 521 approaching Blackwell, circa 1925 with a special to the races at Birmingham's Bromford Bridge, a popular pastime in the '20s and 30s. The L.M.S. coat of arms adorns the cabside — an early practice before lettering on the tender superseded this. *(Real Photographs/D. Sharpe Collection)*

Jinties 1934 and 7107 give a final push to a northbound freight as it nears Blackwell in 1925. Note the squat M.R.-built four-wheeled brake van. *(Real Photographs/D. Sharpe Collection)*

related with electric cables and some of the signals readjust themselves independently of the signalman, as soon as the train has passed.'

There were calling-on arms under both inner and outer 'up' home signals at Blackwell, allowing a train on the Incline to ease tension on the brakes by climbing off the heaviest part of the gradient towards the station. Any following train could be accepted once an express or fast train had passed the half-way point. When a freight train was working over the down track a partly permissive system of working was allowed, bankers being enabled to follow with the pulling-off of a subsidiary arm, this latter fixed on the same post as the Blackwell advanced starter. This showed 'C' or 'W' for clear or warning, according to the position of the freight train on the Incline.

It was noted in May 1933 that 'Big Bertha' was missing from the Incline as she had been taken to Derby for a complete overhaul and minor repairs. While she was away a new coaling stage was opened for refuelling bankers, this being near the south signal box on the up or Newton side of the line. Bunkering had formerly been undertaken on the down or 'works' side of the tracks.

During January 1934 an L.M.S. Beyer-Garratt articulated locomotive (No. 4998) arrived at Bromsgrove to share banking duties with 'Big Bertha', replacing her for a short period while the 'first lady' was again out-shopped. According to local gossip No. 4998 was taking 'Big Bertha's' place for good, but this was denied by an official spokesman for the L.M.S., claiming that the appearance of the Garratt was in the nature of trials only. Two Royal Scot class locomotives also visited the Incline for test and clearance purposes in March 1934, and were the first of their type to appear at Bromsgrove. These were Nos. 6143 and 6145. It was firmly denied that either engine, or others of their class, would work on the Birmingham-Bristol line in regular service and this certainly did not happen for ten years. Engine No. 6143 had recently been overhauled and through the paint shops at Crewe, also changing her name from 'The Royal Mail' to the 'South Staffordshire Regiment'.

During 1936 a larger number of standard Midland Compounds (4-4-0s) began working to Gloucester and Bristol over the Lickey Incline, not being an engine, until then, frequently seen on lines south west of Birmingham, although fair numbers had been diverted to the New Street-Euston services after the grouping. Rebuilt 2P class engines, built by Johnson but rebuilt by Deeley, were much in evidence on slower and shorter trains. Some trains out of New Street were double-headed, the pilot remaining in front until reaching Worcester or even Gloucester. The double-heading duty had frequently been assigned to the so-called Johnson single-wheelers, or Spinners (4-2-2s), at least until the early 1920s. Double-heading was often necessary on the southern exit lines from New Street due to heavy grading from the platform ends; it was also typical M.R. practice as the company was devoted to a policy of smaller engines hauling shorter and more frequent trains than the other main line companies.

Throughout this period the main assistants to 'Big Bertha' were 0-6-0T Jinties, numbered between 7234 and 7238. These five trusty work horses, designed by Johnson, were in daily work over the bank until after nationalisation. Also, Kirtley locomotive No. 20002 was still on her old stamping ground in August 1936, being one of the oldest engines owned by the L.M.S. at that time. On the first of the month she proved her worth by assisting the latest Black Five, 4-6-0, No. 5039, so that ancient and modern were working over the Incline in close harmony. They were in charge of the up 'Pines Express', then one of the crack passenger trains on this part of the system.

One Sunday 7th February 1937 an interesting item of rolling stock passed through Bromsgrove and up the Incline, as part of a special train from Gloucester. This was a private coach ordered by His

Blackwell, July 1929. Midland Railway-built 2-4-0 number 177 on a Sheffield-Bristol express. The first brake van has an L.& S.W.R. look about it, the second is L.N.W.R. By the late 1920s it was not uncommon to see rolling stock from the various pre-grouping companies running in mixed batches the length and breadth of the country. *(Real Photographs/D. Sharpe Collection)*

A pair of 2P 4-4-0s speed through Bromsgrove South with a York-Bristol express in the summer of 1929. Note the Midland Railway signal gantry. *(Real Photographs D. Sharpe Collection)*

Short but sweet! L.M.S days circa 1933, and an 0-6-0 goods engine brings a light semi-fast local up the Lickey Incline banked by a similar locomotive. Note the interesting guards van with a standard ex-Midland Railway covered van to bring up the rear. *(P. Ransome-Wallis/National Railway Museum)*

Highness The Maharajah of Indore, from the Gloucester Railway Wagon and Carriage Company, at a cost of £15,000. The vehicle contained several bedrooms, a drawing room, kitchen-compartment and cocktail bar, with telephonic connections throughout. Because of its awkward size and extra width the coach had to be halted, from time-to-time, to allow the passage of other traffic on parallel tracks, waiting over an hour at Bromsgrove before suitable arrangements could be made for its passage up the Incline. It had already waited an hour in sidings at Stoke Works, about three miles down the line towards Gloucester.

Increased numbers of Class 5's, and later Jubilees, both designed by Stanier, appeared over the Incline from September 1937, eventually becoming some of the better known classes to work trains on the longer distance runs, right until the end of steam workings in the 1960s. These were allocated to Saltley (21A), Gloucester (22B), Bristol (22A) and Derby (17A). Among veterans on the Incline the Johnson bankers, No. 7234-7239, were still active in 1939. They were joined by Nos. 7425 and 7443 of the later series, loaned from Saltley.

While minor repairs could be handled at Bromsgrove during the 1930s, major jobs and overhauls were undertaken at Gloucester, Saltley or Derby. 'Big Bertha' was first sent to Gloucester during the late 1930s. During August and part of September 1939, before and at the outbreak of the Second World War, she passed through Derby Works for re-boilering, whilst later, with the introduction of wartime blackout regulations, there were also modifications to the electric headlamp.

In closing this section on the inter-war period it may be of interest to record an apprehended crime, connected with the Lickey Incline. During the early 1930s a night train from Bristol to Birmingham was known to be carrying large quantities of cigarettes and tobacco. As it ascended the bank thieves hiding on the train, obeying a pre-arranged signal, began heaving crates of cigarettes to their trackside confederates. The receivers were waiting below the wall near Pike's Pool, having a motor van in readiness. The loot was safely collected but someone had warned the police, who were waiting in ambush. The gang was arrested to a man, almost before they knew what was happening.

2290 on Gloucester shed in the 1930s – a rare shot this as repairs were normally carried out at Bromsgrove or heavier work at Derby. The centre wheels are missing so it is in all probability Derby's achilles heel – axle box problems. Perhaps Bromsgrove could not cope and a visit to Derby was out of the question due to the distance involved. Careful nursing to Gloucester shed was the only answer! *(R. H. G. Simpson)*

The Second World War

During the six years from the autumn of 1939, traffic over the Lickey Incline greatly increased, but was subject to an even greater degree of secrecy than during the First World War. Vast amounts of coal and raw materials were being handled, especially by the L.M.S.—then the largest main line railway company in Britain and one of the largest commercial enterprises in the free world. These movements were almost as vital to the war effort as ferrying troops, guns and ammunition. Many new and foreign locomotives passed over the Incline, including a fair number from the Southern Railway—hitherto an almost unknown sight in the West Midland—either as loans or replacements, under a form of national control. Many of the Southern locomotives did not return to home territory until 1945.

A royal train ascended the Incline on 10th April 1940. King George VI and Queen Elizabeth joined it in a siding at Barnt Green shortly after inspecting shadow factories (as they were termed) engaged in secret war work. These factories were camouflaged from enemy bombing planes. The royal party dined and slept on the train for one night only, in Barnt Green, unbeknown to local inhabitants or dozens of commuters then using this stopping place. Before their Majesties dined an informal reception was held on the train, at which both local officials and representatives of the L.M.S. Company were presented. There was a departure for an industrial visit in the Wolverhampton area, at 8.25 the following morning. Breakfast was served while the train was in motion.

During 1940 a searchlight battery of the Royal Artillery, also an anti-aircraft gun emplacement, occupied sites at Finstall Park, near the Incline. It was claimed that the straight almost uninterrupted track, at this location, was an excellent pointer to the City of Birmingham, used as a land mark by numerous German bomber pilots.

To conform to blackout regulations 'Big Bertha' had to have her headlamp partly painted over and covered by a special hood. The rear spectacle plates of the covered tender were also hooded and painted over, while the cab, in common with those of other steam engines, had to be shrouded in opaque, weather-proof blinds, concealing the glow of the firebox. Crews working on the bankers, like other footplatemen, were provided with service pattern steel helmets, gasmasks, anti-gas capes and special boots. On 'Big Bertha' such equipment was carried in specially fitted lockers, while on smaller engines they were kept in tin boxes.

A few attempts were made to bomb the line, the nearest strike being a house opposite Pike's Pool, not far from the railway, in which all the occupants were killed by High Explosive bombs. A freight train, banked by a couple of Jinties, passed nearby while this was happening, but came through unscathed. On reaching the summit the train stopped in Blackwell Station, halted by the air raid sirens. The station master (also a local air raid warden) claimed that a flurry of sparks from one of the bankers, during the ascent, had drawn unwelcome attention to the area.

A slightly later but undated wartime incident, that may have led to far more serious results, concerned a train load of bombs and live ammunition. This was a special freight heading south but halted at the summit for brakes to be pinned-down. On the point of re-starting a coupling link snapped between two wagons, but was noticed in time and a Jinty banker—then awaiting a return trip—connected to rearward stock. In descending the Incline the driver of the train engine noticed that he was pulling too far ahead and slowed to a halt near Finstall Road Bridge. The braking power of the small banker, however, was much less than that of the larger train engine and less under the control of its crew. Unable to stop in time the rear portion of the disconnected train now rammed the stationary forepart. Several wagons reared in the air, one jumping the tracks and almost landing on top of lineside cottages at a much lower level. Little serious damage was done, although the permanent way suffered beyond Finstall Bridge, 400 sleepers having to be replaced.

At this period long and medium size passenger trains were an average of nine coaches long, requiring two bankers, and allowed eight minutes for climbing the bank, end-to-end. Freight trains needed anything up to twenty minutes, managed by the large banker or two or three smaller engines, the latter often Jinties.

Yorkshire coal, as previously stated, was preferred for banking, although not always available in wartime. Apart from other qualities this type of solid fuel came in the right weight, size and shape, being much easier to handle than Welsh coal, a statement with which former Great Western men may not always agree. Bricketts of compressed coal dust were sometimes used, but seem to have been detested by footplatemen, whatever their company, region, or line of duty.

During the greater part of the war Bromsgrove Works was devoted to repairing goods wagons, a vital contribution to the war effort in keeping open lines of supply and communication. A military guard was used to protect the works and for possible security at the foot of the Incline, where sabotage in a complex area of signalling and pointwork could have greatly under-mined the routine of the whole system. These duties were eventually undertaken by the local Home Guard, in which many railwaymen served during off-duty

2290 pictured at Blackwell circa 1933 before returning light engine to Bromsgrove. No L.M.S. identification letters and numbers are visible on the tender or cabside, only the front cast numberplate plus the number 4 shedplate give any means of recognition. *(P. Ransome-Wallis Collection/ National Railway Museum)*

hours. At one time, during the early months of the war, eight serving airmen were attached to the works and running shed, being mainly engaged in engineering duties. Although a short-lived and temporary measure it kept the men occupied and in some sort of training while there was a general shortage of aircraft to fly and service. All the airmen were fully qualified pilots, signed-on for the duration.

On the evening of 18th April 1941, Henry Warren, an instructor at a nearby agricultural college, was returning by car from Bromsgrove when, due to blackout conditions, he mistook his turning and drove on to railway property. The so-called Railway Approach at Bromsgrove resembled an ordinary road and, as a gate leading into the station had been left wide-open, Warren actually drove across the platform, landing on the rails, where his car lodged in a sideways-on position. Fortunately it was fairly late in the evening and few people were anywhere near the station. Apart from slight shock and a badly bruised thigh the driver was unhurt. The car, although badly scratched and having a dented mudguard, was still road-worthy, but had to be cleared with the aid of a banking engine. It could be driven from the station yard and Warren was able to continue his journey without further delay!

Another accident was to occur on the Incline in the spring of 1941. This was a derailment and over-running that delayed traffic in both directions for at least seventeen hours. At 2am on 30th April, a heavy goods train was descending the Incline from Blackwell when, due to broken couplings, two loaded wagons at the rear of the train broke away and rammed the forepart, causing several other vehicles to derail, five of which were badly smashed and their contents destroyed. Permanent way damage extended for three quarters of a mile, some of the debris and damaged sleepers carried as far south as Bromsgrove Station. Fireman Oakley, on the footplate of a following returning banker, sustained painful injuries when flying splinters hit him on face and head—for which he needed first aid treatment. Although two breakdown gangs with steam cranes were called out, almost immediately, it was 6.45pm before normal workings could be resumed. There was a further derailment on the Incline during the summer when an ascending freight train came to grief, but full details were supressed by the censor. Re-railing and general clearance work halted through traffic for several hours. The train had been double-headed by a class 5 locomotive and a class 3F, banked by two Jinties.

In June 1941 'Big Bertha' was back at Derby for overhaul and matched with a new cab-fitted tender. This was a Dubbs pattern Class H type, having a new 2,050 gallon water tank. During the absence of No. 2290 several temporary banking engines were brought in from Worcester Shed, including L.N.E.R. J39 0-6-0 No. 2142, at that time officially loaned to the G.W.R.

Although members of the Royal Scot class, one of the most popular types owned by the L.M.S., were tested on the Incline some years earlier, they did not work over it in regular service until 1942, the third year of the Second World War. The first down the bank, heading a troop train for the Gloucester area, was No. 6161, based at Crewe North Shed. About a month later, in January 1943 No. 6140, from Camden Shed, worked through on several duties between Birmingham, Gloucester and Bristol, and was stabled at Barrow Road Shed, Bristol. Although No. 6140 lost time on an evening turn and was said to have steamed badly, this may have been partly due to poor coal or an inexperienced crew.

In July 1943 'Big Bertha' paid a further visit to Derby and was deprived of her headlamp, for unspecified reasons, although the holding brackets remained intact. She returned to Bromsgrove without one of her more characteristic features, although the lamp was duly restored (again without explanation) several months later.

When the United States of America entered the war, large numbers of steam engines were brought to mainland Britain, either as part of lease-lend programmes or to shunt traffic for the Transportation Corps. in dock areas and at military installations. Two of the first to be seen in the West Midlands were shedded at Gloucester and are reported to have passed through Bromsgrove on several occasions. It may be noted that Gloucester and Sharpness Docks were widely used in preparations for the Allied invasion of Western Europe. The engines concerned were built by Porter of Pittsburgh (U.S.A.). They featured large, fully enclosed cabs fitted with special blackout shutters or blinds. Although numbered 7417 and 7419 on builder's plates, they were given the additional War Department numbers WD1261 and WD1263. Such locomotives always looked much larger than they actually were, as the boilers and upper works had been mounted on high bar-frames, which seemed to give them extra height.

At the close of the war there were many celebrations, especially on V.E. (Victory-in-Europe) Day, to mark the end of fighting in Continental Europe. There were numerous bonfires and fireworks in Bromsgrove, both at an official level and in gardens or backyards along the trackside. Railwaymen celebrated in their own way with loud whistle crows and the placing of numerous fog signals on the Incline, to be exploded by descending bankers as a victory salute.

Chapter 3—The Post War Years

Right away mate! The fireman checks the guard's clearance before 3 cylinder compound 41014 departs from Blackwell with a stopping train to Birmingham New Street in early B.R. days, although L.M.S. is still emblazoned on the tender! The coach is an L.M.S.-built surburban carriage dating from the 1920s. *(D. Sharpe Collection)*

New members of the large Black 5 class of L.M.S. locomotive began to make their appearance on the Incline during 1945. On 21st March an articulated Garratt No. 7868 (Midland type) was noted at King's Norton, later heading towards Bromsgrove for tests on the Incline. This was the first of its type, since the visit of No. 4998 in 1934, to appear at Bromsgrove. Later in the year large numbers of freight and mineral trains were headed over the Incline by Garratts. Several of their number were seen together in Washwood Heath Yard, north of Birmingham, where trains were assembled for freight services in all directions. Before and during the Second World War all Garratts from the north terminated at Washwood Heath or King's Norton.

During the immediate post-war period there were usually eight Jinties, plus 'Big Bertha', on banking duties at Bromsgrove. The Jinties were first brought to Bromsgrove in 1912. In 1947 the running number of 'Big Bertha' was needed for a new class of locomotive and thus changed in September from 2290 to 22290, which remained until a further change in January 1949, this time to British Railways 58100—the final number. An added distinction came when the latest number plate was fitted to the smokebox door, 'Big Bertha' being one of the few engines in the 20000 series to be so numbered. Figures painted on the cabsides, however, were much smaller than in ordinary practice due to the confined space available.

Throughout January and the first two weeks of February 1948, the year of nationalisation, 'Big Bertha' was delayed longer than usual at Derby Works, for overhaul and major repairs. As a rule the men at Derby treated her with both promptitude and affection. Since the early 1920s she was the pride of the stud, known throughout the shops as 'Bertha'. 'Emma', 'Liz', or just 'The Banker'. Any part of exceptional size or quality was known as 'Banker size' or 'Banker type' ... 'Fit for Liz' or 'Good enough for the Banker'.

As from January 1st 1948 the main line route over the Lickey Incline was controlled, as a result of nationalisation, by the London, Midland Region of British Railways.

The ex-L.N.E.R. Garratt came to Bromsgrove on 7th March 1949, working alongside 'Big Bertha' during her final years, but was never in serious competition as anything more than a novelty and engine-spotter's bonus. Designed by Sir Nigel Greseley in 1925, she was the sole representative of her class (the UIs). She had been constructed at Gorton Works, Manchester, by Messrs Beyer, Peacock. Her main assignation had been banking on the Worsborough Incline, near Barnsley, Yorkshire, which carried heavy coal traffic bound for the Woodhead trans-Pennine route. She was displaced from this location when the Manchester, Sheffield and Wath main lines were ear-marked for overhead electrification. Originally numbered 2395 (later becoming 69999 under British Railways ownership) this 2-8-8-2 mammoth was the most powerful steam locomotive in the United Kingdom. Despite her record-breaking potential railwaymen of the London Midland Region took an almost immediate dislike to her, perhaps accounting for her not very welcome period of service on the Incline, although dislike went much deeper than ordinary prejudice. She even arrived at her new depot the wrong way round, her cab-end to the rear, as with more conventional locomotives. Yet for banking on the Lickey Incline it was necessary for a large articulated type to lead from the cab-end or cab-forward position, this being easier for judging distances, buffering-up and affording a better look out on the ascent. She was first sent to Worcester Shed for turning but overlapped the turntable by a considerable amount, a problem that need not have arisen at the cost of a cheap 'phone call. She was finally turned on the King's Norton-Lifford Triangle. For night work on the Incline a large van-type headlamp was attached to the rear of the bunker, operated by batteries.

Number 69999 was certainly no beauty and more than a handful for the men by whom she was driven and maintained. In the words of a Bromsgrove engineman *'She was always short of steam'...'Wouldn't steam properly'* and *'A useless great thing.'* Even during the 1920s and 'thirties there were numerous complaints and criticisms. Her main water supply in Yorkshire was known to be very soft. Lack of protective scaling in turn led to corrosion of surfaces in both tubes and stays. Re-tubing at Doncaster Locomotive Works was necessary as early as July 1926. Corrosion troubles also affected the firebox tube plate in 1927 and the firebox roof stays in 1928. Even when this problem was overcome by the reduction of tube holes and the introduction of chemicals into the water supply, she spent several months shedded at Mexborough where an L.N.E.R. official made a written understatement to the affect that *'they were experiencing some little trouble with the steaming of the Garratt engine...'*

The number 69999 was not adopted until a visit to Doncaster Works in 1948. Numbers then appeared on the buffer beams and tank sides, front and rear. Although the smokebox door was drilled for a cast iron number plate this was never fitted or cast.

In 1949 the Garratt, recommended for a new boiler fitting, but no longer required on Worsborough Bank, was sent to Bromsgrove, where she arrived in the early spring. She was even less satisfied with the water supply in North Worcestershire and developed many other faults and vices. The steam brake was often slow to act so that drivers lost time by approaching with caution, to avoid massive jolting. If there were long

69999 banks a Bristol-York express about 1949. Why the locomotive worked the incline in reverse has never been explained. It may be that due to the tremendous length of the Garratt it was easier for the driver to 'sight' the rear of the train to be banked from the shorter end of the locomotive. Frequent complaints were made that this engine was difficult to gauge in distance. Some guards did comment about severe jolting in the rear coach or van before the train actually commenced its assault on the Incline. *(P. Ransome-Wallis Collection/ National Railway Museum)*

The Garratt descends into a deserted Bromsgrove Station in the high summer of the late 1940s, having rendered assistance to a northbound express. The crew pose for the photographer. *(P. Ransome-Wallis Collection/ National Railway Museum)*

A general view of Bromsgrove South with newly relaid track. A freight train awaits banking assistance whilst three 'Jinty' tanks prepare for action. The driver of the passenger or goods train would 'whistle up' as to the number of bankers required. 69999 has just completed coaling up before its next turn.

(P. Ransome-Wallis Collection/National Railway Museum)

or unexpected delays between duties, cylinders became cool with water collecting in the over-long pipes. Because the reverser tended to slip when working hard, the Garratt was sometimes left behind on the bank while the train she was meant to assist either stalled or pulled-away. On each ascent of the incline she required at least 5cwts more coal than 'Big Bertha' or any other locomotive, which was almost too much of a task for a young, inexperienced fireman. The drivers were unable to help much as their skill and concentration would be needed at the controls. The Garratt always performed better on Wentworth coal, which although risking a firebox full of clinker, provided sufficient quantities of primary air.

She remained at Bromsgrove until early November 1950, returning to Mexborough on the 10th being held in reserve and used for numerous tests and adaptations. Among ideas suggested at different periods, but not accepted, were fitting a revolving coal bunker (which the LMS Garratts had) and the use of a mechanical stoker. There was a further spell as a reserve banker on Worsborough, but she returned to store at Mexborough a year later. A conversion to oil firing was authorised in April 1952 and she was sent to Gorton Works in August for this to be arranged. During this period the battery headlamp was removed and a new lamp fitted, operated by a Stone's generator. There were test runs in traffic over a period of two years, No. 69999 proving shy to steam even in the most favourable conditions. Oil firing did not seem to be the answer to her problems although modifications were made, especially to the oil feed arrangements. It was then decided to return her to Bromsgrove, as a last chance.

She set out from Gorton, light engine, on 29th June, but ran hot by Burton-Upon-Trent where she was delayed overnight. The same happened the day after near Saltley, so she returned to Burton where there was a wheel drop. The affected wheels were removed and taken to Doncaster Works for further attention. She returned from Burton on 7th July, arriving safely in Bromsgrove later the same day. The second stint on the Lickey Incline commenced the next morning. Her fuel tank now had a capacity of 5,000 gallons of oil, sufficient for 24 hours steady work. This in itself caused problems as, when steam was shut off, at the top of the Incline, clouds of black, oily smoke drifted across a nearby golf course and neighbouring gardens. As may be imagined, there were numerous complaints from playing members, to say nothing of local householders. On 13th September 1955 she was again returned to store, spotted at Bromsgrove Shed...'*cold and uncared for, while covered in layers of dirt and grime.*' A few days later she went to Burton, continuing to temporary storage at Gorton—after a short delay—via Chesterfield and Guide Bridge. She never worked again in revenue service and was condemned in November 1955, last seen on a dead-end siding at Doncaster Works shortly after New Year's Day 1956, awaiting the torch.

On 16th April 1955 a special rail tour ran from Paddington to Bristol and Birmingham, known as 'The Lickey Limited'. The train was hauled by Castle class locomotive No. 7017 'G. J. Churchward'. being only the second time this type of Great Western Railway designed locomotive had climbed the bank, the first being in January of the same year when 'Lydford Castle' passed through Bromsgrove to King's Norton on clearance tests. 'G. J. Churchward' had recently served on royal train duties, having brought Queen Elizabeth II through the West Midlands

Love's Labour Lost! U1 69999 awaits scrapping at Doncaster Works in 1956. This locomotive had been converted to oil burning the previous year. After further trials, British Railways decided to withdraw the monster Garratt as the running costs proved totally uneconomic.
Regarding 69999's spell on the Lickey, rumours were rife at Bromsgrove m.p.d. that every effort was made by the crews to ensure the engine remained unpopular. If it had proved a success, in all probability, at least three of the 'Jinty's' crews would have become 'surplus to requirements'! *(Derek Harrison Collection)*

to Windsor, after a tour of industrial Lancashire. The driver-in-charge was a Mr Potter of Old Oak Common, by whom the royal train had been driven. The 'Lickey Limited' was fully subscribed with 420 passengers, enjoying buffet car service and the amenities of a spacious, beaver-tailed observation car. The latter item of stock was ex-L.N.E.R., a type seeming more at home behind an A4 Gresley Pacific than a Western Region Castle. There were numerous onlookers, railway photographers and train spotters at vantage points along the Incline, side lanes jammed with private cars of all makes and sizes, including a vintage Rolls Royce. For the ascent of the Incline 'Big Bertha' tramped at the rear, buffered-up to the observation car.

Several months after banking the 'Lickey Limited' 'Big Bertha' was back at Derby for boiler checks. During her absence No. 69999 returned, while the far more recent Standard class 9F 2-10-0 No. 92008, recommended as an assistant banker, came on shed at Bromsgrove. There were eventually several 9Fs to work on the Incline—at different times—for periods ranging from a few days to a few years, not always as spectacular or picturesque as 'Big Bertha' but performing yeoman service. Railwaymen at Bromsgrove now began to surmise, for the second time, that 'Big Bertha' would not return to them. There was even talk of a farewell dinner in her honour while premature obituaries appeared in several local newspapers. She did re-appear for normal duties at the end of August, but reprieve was short-lived as she was finally withdrawn from service in May 1956. Eight months later she went to the torch, although her multiple-cylinder block was preserved as a relic of special interest. She had spent most of her working life in North Worcestershire, the only interludes being trips to outside repair shops and a short trial stint on Toton-Brent coal

Rare visitor! Western 'Castle' class 4-6-0 number 7017 'G. J. Churchward' storms up the Incline with 'The Lickey Limited' (Paddington-Bristol-Birmingham New Street-Paddington) excursion organised by Ian Allan, the railway publishers. The date, 16th April 1955. The 'Castle' came off the train at King's Norton due to severe restrictions on the curve towards Bournville and proceeded via the Camp Hill line to rejoin the train for the return run to Paddington. *(G. F. Bannister)*

With a little help from a friend. 58100 'Big Bertha' banks 'The Lickey Limited' towards Blackwell on 16th April 1955. Note the 'beaver-tail' observation car built for the pre-World War II 'Coronation' express introduced by the L.N.E.R. *(Roger Carpenter Collection)*

trains whilst stabled at Wellingborough.

Shortly before withdrawal she had a final chance to prove herself when called out to rescue a stalled train of nearly fifty wagons, already half-way up the Incline, headed by Garratt (Midland type) No. 47972 and assisted by ex-L.N.E.R. Garratt No. 69999. 'Big Bertha' managed to start them rolling! 'Bertha's' last banking trip over the Incline was on Sunday night 6th May, when she gave assistance to the up mail and then went 'on shed' about 1am on Monday morning, leaving for Derby at 5am. She ran light engine to her final destination, in the care of Deputy Shed Master L. Rudge, limited to a speed of 25mph throughout. During her final period of service she operated with 19in. bufferheads, while her tender had three-link coupling of the plain type.

On two separate Sundays in March 1955 there were numerous trials on the Incline related to future operational needs. Up to that time it had been a rule, especially under the L.M.S., that train weights above 90 tons needed to be banked or assisted. For the trials three ascents were made by class 5 4-6-0 No. 44776 (6th March) and three by Jubilee No. 45554, also a 4-6-0, a week later. All trailing and trap points were locked throughout the duration of the trials. Number 44776 headed a seven coach train, including a dynamometer car, and began its climb in mixed weather with side winds and heavy gusts of sleet, commencing on the stroke of 11.30am. It started at Stoke Works Station, keeping to the 30mph speed limit through Bromsgrove, and climbed to the summit without incident or delay. Three stops and restarts were then required; one mid-incline, one at milepost 55 and the other opposite Blackwell distant signal. There was violent slipping at the first attempt but 44776 managed an excellent start at the second. The Jubilee, with a load of 252

Early days with a 9F 2-10-0 as banker. 92008 prepares to bank the 4.45pm Bristol Temple Meads-York express on 17th August 1955. Bromsgrove Station. *(Michael Mensing)*

Rivals! A rare photograph taken near to Vigo of 69999 and 58100, both assisting a northbound freight towards Blackwell. Is this the event described in the text? If not one wonders why the two largest, most powerful locomotives in the area were working together. The sounds from these monsters must have been deafening! *(P. Ransome-Wallis Collection/National Railway Museum)*

'Big Bertha' — The End
Derby Works circa 1957 with 58100 being lifted clear of her running wheels. Note the flange score marks on the frames towards the rear. The horn guides appear badly worn too. For this unique locomotive it was indeed a terminal case.
(British Rail, National Railway Museum/ D. Sharpe Collection)

The boiler being raised off the frames after the cab was removed. Once the major components had been dealt with it was easier to cut up the chassis into furnace-length components. Plans were made to preserve the cylinders for eventual display in a museum but not proceeded with. *(British Rail National Railway Museum/D. Sharpe Collection)*

tons, including an extra coach, made four ineffectual attempts to start from the first mid-way position, returning to the foot of the Incline in reverse. It started successfully from the second position and fairly stormed up the rest of the Incline. Both engines managed a standing start and unassisted through trip from the up platform of Bromsgrove Station without apparent difficulty.

During the first month of 1956 news leaked that the Birmingham-Bristol route over the Incline had been passed for both Britannia Pacifics and the Fell experimental diesel locomotive. The Britannia 'Tennyson', No. 70032, came on 7th April, heading an excursion connected with a schoolboy's Rugby International and also handled the return working. The Fell locomotive, however, did not make a showing either at Bromsgrove or on the Incline.

Early in 1956 a deputation of local housewives, living near Bromsgrove Station and at the foot of the Incline, made complaints to the Bromsgrove Sanitary Inspector concerning pollution caused by bankers, especially when two or more were working together in tandem. It claimed that on such occasions the surrounding countryside was hidden under a pall of dirty smoke, it being unwise to expose newly washed clothes in lineside gardens. Apologies were made, also promises, but no direct action was taken.

Towards the end of June 1956 there were clearance tests over the Incline for several ex-G.W.R. engines, including 0-6-0PT No. 9432, which appears to have fouled lineside objects and tunnel mouths north of the Incline. On 6th December ex-G.W.R. 4-6-0 No. 7819 'Hinton Manor' headed a train of twenty hopper wagons through Bromsgrove, making for Barnt Green. This locomotive was a novice on the Lickey route, in common with the rest of her class. (She was later purchased and preserved to run on the privately-owned Severn Valley Railway).

By the end of 1956 most of the ex-Midland type locomotive had departed from the Incline, No. 47576 however, remaining in residence until 1958, while others made intermittent appearances. Bromsgrove was for long regarded as a sub-shed of Worcester (Midland Railway code 4) until 1922. Under the L.M.S. it changed to a sub-shed of Saltley and was known as 21C, although Saltley was 21A. By 1958 it was changed to 85F again under the control of Worcester.

During 1957 a number of local passenger train workings in the West Midlands were taken over by new diesel-multiple units. It was recorded that on 2nd June two d.m.u.s were coupled together for a day excursion trip to Weston-super-Mare, the first working of this type from Birmingham over the Incline. Before long, however, d.m.u.s were common-place for many duties. The last steam-hauled service, for example, between New Street and Great Malvern, ran as the 9.24am during July 1960. This, as usual, consisted of six coaches of ex-L.M.S. stock, hauled by a mixed bag of motive power, usually from Saltley Shed.

From January 1957 a number of 9400 class 0-6-0PTs began to appear for work on the Incline, transferred from the Western Region via Worcester. These were Nos. 8400-6, fitted with L.M.S.-type lamp brackets and L.M.S. pattern continuous blow-down valves. They operated on normal duties in pairs, threes and fours-according to conditions and requirements.

As from February 1958 the Birmingham and Bristol line, south of milepost 52½ near Blackwell, at the summit of the Incline, was transferred from the London Midland Region to the Western Region of British Railways. A greater number of ex-G.W.R. locomotives and items of rolling stock were soon spotted in the area and no longer a curiosity on the Incline. An ex-G.W.R. 2-8-0T No. 5226 came to Bromsgrove Shed in 1958 and was given extensive trials for the Incline. It remained for two years and gave all-round satisfaction, but was merely on extended loan from Ebbw Junction Shed, Newport, by whom she was again required in 1960. An ex-G.W.R. 2-8-2T, No. 7235, was later brought in for banking tests but failed, due to clearance difficulties in Bromsgrove Station.

During 1959 there were reports of several ex-L.N.E.R. engines heading trains through Bromsgrove. This trend began early in the year with a few class B1 4-6-0s spotted on the Incline. There were also the first recorded visits made by V2 locomotives, normally banned south west of Birmingham, on York-Bristol trains. The first visit was made by No. 60954 (York 50A), which changed crews at New Street, Birmingham, and should have returned back from this point, but was worked forward to Temple Meads, Bristol. At Bristol she was met by an angry reception committee and impounded by the Western Region Authorities as unsuitable for the route. She was sent back the next day, under strict supervision and speed restrictions, spending the night of the 17th November at Bromsgrove Shed. A second V2, also on a York-Bristol train, slipped through the mesh of red tape a few days later and was impounded at Barrow Road Shed, Bristol from the 12th December to 18th of that month, eventually returning light engine.

On 9th December 1959 the prototype of the Royal Scot class No. 46100, worked a passenger train from Bristol to Sheffield, having been transferred with others of this class over the Incline, before their final demise in 1963.

The 'Royal Scots' and Patriots came to the route due to the increasing dieselisation of the Euston-Crewe-Carlisle main line, being re-allocated to Millhouse Shed, Sheffield. They included 45514 'Holyhead', 45536 'Private W. Wood V.C.', 46131 'The Royal Warwickshire Regiment' (named at New Street Station, Birmingham in 1938), 46148 'The Manchester Regiment', 46151 'The Royal Horse Guardsman' and 46164 'The Artists Rifleman'. All the above named worked over the Lickey Incline regularly.

Earlier in the year (1959) Mr J. Hudson, Chairman of the Bromsgrove

Post-War Pannier 0-6-0PT number 8402 and another class member pound their way out of Bromsgrove on a goods train banking duty on 1st June 1959. *(John Edgington)*

4200 class 2-8-0 tank number 5226, built by the Great Western Railway in the 1920s, on banking duties 26th September 1959. Viewed from the brake van verandah as the locomotive falls away from the rear of the freight train to run via the crossover in Blackwell Station and back down the Incline to Bromsgrove. *(H. C. Casserley)*

Restored. The famous gravestones in St. John's Churchyard, Bromsgrove. *(British Rail)*

Ex-L.N.E.R. B1 4-6-0 number 61041 on a relief holiday express pauses at Bromsgrove before ascending the Lickey Incline on 6th August 1960. *(John Edgington)*

Urban District Council, was in contact with British Railways concerning the sad condition of the gravestones recording the death of Scaife and Rutherford killed when a locomotive exploded after duty on the Incline. The stones, in Bromsgrove churchyard, had been neglected and vandalised, deteriorating to an unsightly condition. British Railways, having taken full responsibility for the site, promised to make a thorough survey and renovations. Mr George Dow, a representative of B.R. from New Street, Birmingham, visited the graves on 13th October and repairs were soon put in hand.

During 1961 there was a steady increase in Royal Scot class workings, several of these engines being re-allocated to Saltley and Derby while others went to Trafford Park (Manchester), along with a number of rebuilt Patriots (Baby Scots), mainly for fast passenger trains from northern centres to Bristol. Main line diesel locomotives also started appearing for crew training on these routes, including the Incline, early in 1961. They were first seen on active service, dealing with summer holiday trains from north of Derby. On 24th June Sulzer type 4 No. D37 was seen moving up the Incline with a Bristol-Bradford train, banked by an ex-G.W.R. pannier tank from Worcester, No. 9486. The long reign of steam was gradually drawing to its close.

An article in one of the leading railway magazines during 1961 revealed that the Western Region of B.R. was eager to dispense with the cost and trouble of banking engines within their territory. Although both diesel and steam trains continued to be banked on the Lickey Incline it was obvious that most diesels could look after themselves. With this in mind an official test was held on 20th September, B.R. Sulzer type 4 No. D40 heading a sixteen-coach train, including a dynamometer car, most of the vehicles being empty stock. Eight bogies were added to simulate the weight of missing passengers and their luggage. This was to be a test not only of hauling power but the strength and reliability of couplings, so the resident 9F banker, No. 92234, was ordered to bring up the rear, at a respectful distance, checking possible runaways. The Incline was attacked from both standing and flying starts, D40 emerging with great credit, the average time of ascent being seven minutes. The next day similar tests were made with assorted freight stock.

On 16th February there were notable tests laid on for the Brush built diesel-electric experimental prototype D0280 'Falcon'. This locomotive made a successful climb out of Bromsgrove Station with eighteen coaches and the Western Region dynamometer car, totalling a weight of 638 tons at the drawbar. Tests were repeated several times in succession, then with freight stock—the latter being 618 tons. The prototype diesel hydraulic locomotive 'Lion' D0260, was on similar trials in May and June, able to move 500 tons from dead stops on the Incline—unaided—with sixteen to twenty items of coaching stock. This latter locomotive was built as a joint venture by the Birmingham Railway, Carriage and Wagon Company Limited, Sulzer Bros. Limited and Associated Electrical Industries Limited.

As a result of various earlier tests a decision was made that banking over the Lickey Incline was no longer compulsory in all cases. The first regular service train to complete the ascent, unbanked, after this ruling, was a Bristol-Newcastle fast passenger train. This was made up of nine coaches hauled by a Peak (B.R. Sulzer type 4) diesel-electric locomotive. It is worth noting that during the heyday of steam several local passenger trains, with only two or three coaches managed to struggle over the Incline unaided, especially when there was a failure of the banker. Such attempts, however, were unofficial and strongly discouraged!

In September 1962 the named train

Reflecting the increase in workings by the Royal Scot class a grimy 46151 'The Royal Horse Guardsman' enters Bromsgrove Station, along the down through road, with a West Country holiday relief in the summer of 1961. The trainspotters on the up platform would certainly end up with a page full of numbers on any summer Saturday!

(B. S. Moone)

'The Cornishman' was transferred to pick up/set down at New Street (Birmingham) instead of Snow Hill, effective from the beginning of the winter timetables (9th September). This had formerly been routed from Wolverhampton low level via Snow Hill, Stratford-upon-Avon and Cheltenham to Penzance, but later traversed the Lickey Incline via New Street and Bromsgrove to Cheltenham etc. At about the same time the 'Pines Express', having used the Midland route via Bristol, Bath and the Somerset & Dorset line to Bournemouth since 1927, was re-routed from the north of England via Oxford, Reading and Basingstoke.

During the winter of 1963, due to severe weather conditions, many diesel locos were frozen-up and steam locomotives returned from store to work their rosters. For the remainder of the season a number of Jubilees worked from Burton Shed (16F), while an unexpected visitor to the area was Britannia class 4-6-2 No. 70018 'Flying Dutchman'.

The first production version of the Brush diesel electric type 4, No. D1500, built at Loughborough, underwent bank tests on the Lickey Incline in February 1963, making unassisted stops and starts with a train of nineteen coaches and a dynamometer car, total weight being 690 tons. The 9F banker, as on other diesel tests, trailed at a safe distance throughout.

About this time it was decided to close Bromsgrove Works, after a short running-down period. Total closure was to take place by the autumn of 1964, this being part of the Beeching plan to cut-down and economise on all aspects of railway working throughout the country, eventually leading to the closure of over 200 stations and many branch lines. Twenty-one employees at Bromsgrove were given immediate notice.

On Sunday 14th June 1964 there was a rail tour passing over the Lickey Incline en-route from Bristol Temple Meads to Derby Works, organised and sponsored by the Warwickshire Railway Society. The locomotive in charge, representing a rare visitor to the West Midlands, was the Southern Region Battle of Britain class 4-6-2 No. 34079 '141 Squadron'. At this period steam and diesel locomotives were still being used over the Incline in equal numbers, with many interesting visitors from outside areas. Although Royal Scots and Patriots had been dropped, there were still a few ex-L.N.E.R. B1s and a number of ex-G.W.R. engines, especially from the Worcester area.

The first diesel banker, an English Electric type 3 (later class 37), No. D6938, arrived at Bromsgrove Shed for crew training on 2nd July, full dieselisation of this service was then anticipated from September 1964.

On 12th July 1964 a loco-spotter's special, working via the Incline from Leeds to Cardiff, was headed by ex-L.N.E.R. 'A1' class locomotive No. 60114 'W. P. Allen', the only representative of the A1s to traverse this route. Unfortunately it failed at Worcester with a damaged connecting rod and was detained for several weeks at the local shed, later returning under its own power light engine.

Brush type 4 No. D1733, smart in a blue experimental livery, was banked up the Lickey Incline with a heavy freight, on 15th July 1964, a day further brightened by the appearance of an ex-G.W.R. Grange class engine on a north-bound mixed freight, and Britannia class No. 70004 'William Shakespeare' also in charge of freight, banked by D6938.

Early in 1964 local newspapers announced a recommendation, made by the Birmingham division of British Rail, that many local passenger services, then costing over £750,000 per annum, should not merely be pruned but terminated. Bromsgrove, Barnt Green and other nearby stations were to fall under the axe, which caused many protests throughout the area. Bromsgrove Urban District Council gave their full support to these complaints, holding a special meeting on 4th September.

The later part of the year was dominated by the complete closure of Bromsgrove Works and the adjacent Locomotive Shed, both of which had been open for over 120 years. Most of the machinery and fittings in the remaining wagon repair shops were dismantled or sold as they stood, mainly for scrap. At its peak the repair plant had handled 200 items of rolling stock per week, employing 500 men, although numbers had dwindled to 67 by the end of 1964. Most of the staff and employees were made redundant from July onwards, the last day of operations being 2nd October. The site covered by the complex was 16½ acres and was later acquired for commercial development. However, most of the buildings remained intact but unused, eventually sprouting weeds for seventeen years before being demolished in 1981. The status of the motive power department, at the closure of the works, was reduced from a sub-shed of Worcester to a mere stabling point.

Steam banking engines were reallocated and withdrawn from Bromsgrove as from 1st September 1964. There was, however, a limited amount of steam banking on the Incline until late 1965, by engines drawn from outside the area. By 7th November 1964 the regular English Electric type 3s Nos. D6922, D6938 and D6939. Number D6922 was deputising for a fellow class member then under repair, but landed in trouble by over-running and demolishing the buffer stops on the banking spur.

It was the aim of the operating authorities to eliminate steam workings in the region by September 1965, but this was impossible until much later. Steam was, however, officially withdrawn from regular freight workings between Birmingham and Bristol as from 1st March 1965.

Although British Rail had announced the closure of several local stations during the winter and spring of 1965, the strength of opposition was so great, from both private individuals and local government, that a stay of execution was granted. A public meeting booked at the Grand Hotel, Birmingham, for the evening of 16th March, was fully attended, at which objections—from all sides—were heard and discussed. There were over 300 objectors in person, with 948 written statements and depositions submitted. The hearing lasted five and a half hours in full session. Evidence against the closures, supported by local M.P.s, the Worcestershire County Council and many local councils, centred on poor alternative bus services, effects on trade and the congestion of local roads. These protests appeared to have the desired effect—but with limitations—as only two of several stations were closed outright (Blackwell and Stoke Works), whilst remaining stations were reduced to the status of unstaffed halts. For a time, however, Bromsgrove retained a small but fully staffed booking office.

There was an accident at the foot of the Incline on Saturday 16th January 1965, at 9.15pm. This involved a steam-hauled goods train, banked by two type 3 diesels. Shortly after starting from Bromsgrove a mineral wagon derailed and over-turned, affecting at least seven of the following wagons. As a result both lines were blocked for 24 hours. Distorted track had to be lifted and replaced for well over 100 yards, while wreckage was removed by steam crane. There were no human casualties.

Although diesel power now dominated the Incline, a few steam hauled workings continued until well into 1966, many of these being substitutions for failed diesels. From early March all but two former steam turns between Birmingham and Bristol were handled by Brush type 4s. A few steam locos however, appeared on specials running for enthusiasts and train spotters. Warship class diesel-hydraulic No. D870 'Zulu' was noted on certain northbound workings of the 'Devonian' from the middle of March.

British Railways built Castle class No. 7029 'Clun Castle', later acquired by the Birmingham Railway Museum, but then still working for British Rail, headed an enthusiast's special from Birmingham Snow Hill to Swindon Works, on Saturday 2nd April 1966. This returned via Oxford, Worcester and the Lickey Incline, although then unable to reach New Street, its planned place of termination, as a result of clearance problems.

The famous 'Flying Scotsman', ex-L.N.E.R. A3 Pacific No. 4472, also hauled a rail-spotter's special to Swindon the following month for a conducted tour of the railway works. This started from Nottingham (Midland) Station, making a round trip via London (allowing time for a short visit to the Museum of British Transport at Clapham), returning through Gloucester and over the Lickey Incline. It was noted that the 'Scotsman', despite the impatience of the banker's

When things go wrong! On 16th January 1965 a northbound freight was about to be banked out of Bromsgrove by two type 3 diesel locomotives, including D6938. These two views show the result of over-enthusiasm by the banker's train crew. A 13-ton wooden open wagon suffered worst of all, being virtually reduced to matchwood! Class 8F 2-8-0 number 48027 was despatched from Saltley to deal with the ensuing blockage. The line remained out of commission for quite some time. All trains were diverted via Kidderminster and Stourbridge Junction until traffic was once again restored 'Over the Lickey'. *(Harold Thompson)*

crew, took on plenty of water in Bromsgrove and climbed the Incline in six minutes, its vocal effects captured by many enthusiasts with tape recorders, both on the train and at the lineside.

The month of June 1966 saw the official end of steam for banking, latterly using, very rarely, a Worcester-based engine of G.W. or W.R. origins. Most duties at Bromsgrove, from that month forward, including shunting, were the preserve of the English Electric type 3 diesels, later assisted and replaced, for a short spell, by diesel-hydraulics.

Although steam remained on parts of the national system for another two years, it virtually disappeared in the West Midlands during 1966, apart from the movements of preserved stock, industrial or privately-owned locomotives. Steam to be observed during the first three months was mainly represented by Black 5s, Stanier 8Fs and a few Standard class 9Fs all on freight, ballast or engineering duties. One of the Black 5s working as late as 16/17th July on a Wareham-Ellesmere Port oil train, developed a fault while passing through Worcester and had to return to its home base at Oxley Shed, Wolverhampton (2B) via Stourbridge Junction.

Extensive banking tests were given to Hymek diesel-hydraulic locomotives during the latter part of July 1966, it being anticipated that such machines would share duties on the Incline in less than twelve months from then.

During the autumn of 1966 work commenced on a large oil and petrol storage depot at the southern approaches to Bromsgrove Station, with extensive sidings for rail traffic. It could be further entered by road from Station Approach. The first delivery was a 500 ton block movement from Plymouth, via Exeter and Bristol. Most of this rail traffic, however, now arrives from the northerly direction, over the down line through the station.

The R.C.T.S. 'East Midlander' hauled by L.N.E.R. Pacific number 4472 'Flying Scotsman' accelerates away from Blackwell summit on 29th May 1965. English Electric 1750HP Diesel Electric number D6943 falls away from the rear of the train after rendering banking assistance. *(Michael Mensing)*

The oil and petrol storage depot stands to the south of Bromsgrove Station and can be clearly seen, with its rail connection, in this view of a three-car class 116 suburban diesel multiple unit forming the 09.26 Great Malvern to Birmingham New Street on 30th December 1985. A good complement of intending passengers wait for it to stop. *(C. J. Tuffs)*

After the final elimination of regular steam workings over the Lickey Incline, most traffic between Bristol, Gloucester and the north was handled by Sulzer and Brush type 4s (later classes 45/46 and 47). The English Electric type 3 (later class 37) bankers were replaced by Hymeks in November 1967, the first locomotive of this type—for regular duties—being D7011. It may be noted in this connection that the first gears of all Hymeks working the Incline were permanently disengaged. On the whole, they were not well suited for this type of work, resulting in many failures and delays. About the same period Warship class diesel-hydraulic No. D855 'Triumph' was spotted between Bristol and Birmingham (New Street) on crew training duties.

On 24th November 1967 a train of empty coal wagons became derailed near the foot of the Incline, on the down side, in the area of Bromsgrove Station, causing considerable damage to the platform and platform facings. Over 40 men and a steam crane were needed for clearing the debris that affected over 200 yards of track. The line was closed to all traffic for twenty hours. Partly as a result of the damage, Bromsgrove Station was entirely rebuilt during a three month period between April and June 1968. This also led to the demolition of the passenger footbridge and the removal of the entire down platform. A short temporary platform was used during April and May, the new structure being ready by mid-June.

During their relatively short period of banking duties at Bromsgrove two 'Hymeks' — 7022 and 7021 — are depicted thrashing away at the rear of the 8M63 7.30pm Severn Tunnel Junction-Washwood Heath freight service, with class 45 number 119 at the head. Unlike the present day class 37s, 'Hymeks' were not fitted with multiple working cables, so a train crew would be on duty in each forward facing cab. *(G. F. Bannister)*

A further derailment and stoppage occured on the Incline during Monday 30th June 1968, when a mixed freight with several car-transporter vehicles came to grief. Couplings snapped during the descent, four wagons being rammed and over-run by the rearward section of the train. One of the cars carried was damaged beyond repair. The driver in charge of the train ran down the Incline waving his arms and managed to halt an ascending passenger train before this could be involved. Both lines were blocked until 7pm the following evening, after twenty hours delay.

In January 1972 one of the first notable strangers over the Incline for a while was an eight-car Pullman set, chartered for a Bristol-Leeds football special. This was now an unusual sight in the West Midlands, as the Birmingham Pullman (Wolverhampton-Snow Hill-Paddington) had long ceased running.

By this time a fair number of Western class diesel-hydraulics had been evident between Birmingham (New Street) and destinations in the south west of England/South Wales. However, they began to disappear during the mid 'seventies, their last working over the Incline being on 11th February with an Exeter to York and return enthusiasts charter, appropriately organised by Gloucester based F. & W. Railtours, D1023 'Western Fusilier' featured.

The Hymek locos began to be withdrawn, after their short reign at Bromsgrove, from October 1972. The first of their replacements was a class 37, No. D6910 (later re-numbered 37 210), originally from Landore Depot, Swansea (87A).

Early in June 1977 a new bridge was constructed under the lower part of the Incline, spanning the Bromsgrove-Redditch dual carriageway. Ready-made pre-cast beam sections were installed, with the earth and clay of the Incline dug away from underneath, allowing minimum interruption to either rail services or general construction work.

Number 7029 'Clun Castle', by now a preserved locomotive based at the Tyseley Depot of the Birmingham Railway Museum, passed southwards through Bromsgrove on 6th October 1978, en-route for Hereford to work a steam special. On the same day preserved Western diesel-hydraulic No. D1048 'Western Lady' was hauled dead up the Incline after repairs and overhaul at Swindon Works. 'Clun Castle' returned to its home base, via the Incline, two days later.

In 1981 a regular service of InterCity 125 High Speed Trains commenced running up and down the Lickey Incline as part of re-cast north-east/south-west services, often surmounting the summit at speeds in excess of 60mph. The first, and only, visit of a British Rail Eastern Region Deltic diesel locomotive also took place, on Saturday 28th November 1981 in the evening darkness, headed by 55022 'Royal Scots Grey', with a return enthusiasts charter from York to London Paddington.

Due to falling freight traffic over the Incline bankers were withdrawn at the weekends as from 7th March 1981. The two engines on duty at this time worked back to Gloucester Shed (their home base) at mid-day Saturday, reporting back for further duties at Bromsgrove on Sunday evening. From 1985, at weekends, the two class 37 locomotives, having operated in multiple Monday to Friday, were separated, with one returning to Worcester and the other to Gloucester. The Worcester engine would bank the nightly Bristol-Scottish sleeper service on its own, proceeding then light engine to New Street Station to pick up the Sunday morning newspaper train bound for Worcester. Bromsgrove remained a booking-on point for local drivers and assistants, until the responsibility was taken over by Gloucester Depot based men.

Steam again returned to Bromsgrove on 11th April 1981, when the preserved Ivatt 2-6-0 No. 43106 ascended the Incline, light engine, en-route from Kidderminster to Tyseley and the Birmingham Railway Museum, from which it was to work an enthusiast's special to Didcot.

On Saturday 18th July 1981 a Sulzer type 4 diesel No. 45052, caught fire during the descent of the Incline. This was at the height of summer traffic, causing considerable discomfort and inconvenience to the travelling public. The locomotive concerned was heading the 8.37am from Leeds to Paignton (reporting number 1V78). It came down the bank with brakes applied, a signal light at Carrington's Bridge appearing to be against it, and passed under the bridge at less than walking pace. According to observers, flames and smoke began to appear at the rear-end. Just beyond the bridge the locomotive made strange 'whooping' noises. The driver soon brought his charge to a halt, dismounting to uncouple and moving the locomotive away from the train at great personal risk. He then attempted to fight the flames with two fire extinguishers normally carried in the cabs, but these were not enough and were soon exhausted. As the extinguishers were both empty long before the fire could be brought under control it was necessary to summon local fire engines by telephone. The brigade, however, found it extremely difficult to approach the disaster, through twisting lanes and over undulating fields. It was more than an hour before the fire could be put out and the locomotive towed to an empty siding near Stoke Works, having caused a 'train-jam' to its rear. Wrong line workings were eventually allowed and the situation eased. The guard of the stranded train, on making his rounds, discovered that a young couple had urgent reasons to make a connection for South Wales and was able to stop another southbound train so that they could transfer to a different service without delay. A relief locomotive, No. 47332, eventually came through from Gloucester Shed, there being an enforced delay of two hours 23 minutes to the hapless Devon bound express.

As a result of the increased workings and speeds by High Speed Trains, certain improvements had to be made to the permanent way on and near the Incline, especially between Blackwell and Barnt Green. Most of this was done on the Saturday and Sunday of the weekend commencing 21st March 1982; other similar work continued until 4th April. This latter involved relaying junction points at Barnt Green and the construction of both retaining walls and sleeper barriers at Blackwell, to prevent threatened landslips. Certain elderly lamp posts, marking a fenced-off pedestrian way at the summit, were removed as was the instruction board at the northern approaches to the summit which read:

ALL GOODS TRAINS MUST STOP AT THIS POINT TO HAVE THEIR BRAKES PINNED DOWN.

Steam locomotive ex-L.M.S. Jubilee 4-6-0 No. 5690 'Leander', passed up the Incline at 4.40pm on 13th March 1982, en-route for wheel turning at British Rail's Tyseley Depot. This movement was from Bridgnorth on the preserved Severn Valley Railway. 'Leander' returned on Friday 26th March, running tender first to Droitwich, and was greeted by large crowds of enthusiasts at various points along the route. This locomotive had previously worked many services over the Incline between Birmingham, Gloucester and Bristol in the final years of main line steam operations, being based at Bristol's Barrow Road Shed for some time.

'Clun Castle' passed again through Bromsgrove and over the Incline, running between Tyseley and the Severn Valley Railway, on 19th June 1983. She returned, tender-first, the following weekend. On both occasions she was in charge of an official Western Region Inspection Saloon, also based on Tyseley.

Throughout the period from the late 1960s the operation of local services have been under promise or threat of closure, this echoed in the misfortunes of Bromsgrove Station, even though events were claimed to have stablised through the support from local authorities. For example, in the summer timetables for 1986 trains calling at Bromsgrove, once a stopping place for important services in both directions, were reduced to three per day, with no Sunday service.

It is important to emphasise, however, that the Lickey Incline itself is a feature of one of the main cross-country traffic routes in mainland Britain. It is thus unlikely to be relinquished while there is still a viable rail network at national level.

Postcript

Bromsgrove Station is shortly to be revitalised as the District Council is to spend £10,000 on extra car parking. Talk of turning the run-down single platform station into a 'Park and Ride' type establishment has been the subject of discussion between the local authorities and British Rail for some time.

A cash boost was initially offered to Hereford and Worcester County Council for a second platform at Bromsgrove Station. Now the county's public transport sub-committee has stated it will pay the full £60,000 for the platform so that the district money can be invested into a more substantial car park to serve the increased populace in the surrounding area.

In all probability a properly staffed railway station will occupy the site at Bromsgrove once again.

The end of February 1977 saw the final withdrawal of the remaining class 52 diesel-hydraulic locomotives from British Rail stock. A number of 'farewell' special trains were organised taking these 'Western' named locomotives to various parts of the railway system where their appearance would turn a head or two!
One of these sell-out tours was the 'Western Finale', organised by the Gloucestershire-based F. & W. Railtours, and ran from Exeter to York and return. D1023 'Western Fusilier' headed the eleven coach train and is seen here foot-sure on the Incline in damp conditions on Saturday 12th February 1977. After ploughing into a herd of cows a couple of hours earlier at Charfield, damaging brake pipes, the late-running tour is back on the move thanks to a suitably filed ten-pence piece acting as a plug! *(Andrew Bannister)*

Chapter 4—General Workings and Banking Practice

The Lickey Incline comprises over two miles of straight, undeviating, double track. It begins at the northern throat of Bromsgrove Station and climbs north east to the site of Blackwell Station (closed 1966), through an area known as Finstall, skirting the village of Lickey End to the west of the line. Bromsgrove Station, now a halt, is actually in the suburb known as Aston Fields.

The only comparable incline on a main line still worked by British Rail, is Dainton Bank in south Devon also within the Western Region, between Newton Abbot and Totnes. At Dainton, approaches on the west side of the summit have short rises of 1 in 36 and 1 in 37, but the ruling gradient is in the 60s. Other banks, long famed in the railway history of the British Isles, such as Shap and Beattock, are comparitively long drags rather than dramatic rises. With the Lickey Incline there are no sudden falls to compliment the rises and the only alternative to reaching the summit would have been by roundabout routes (as was first contemplated), or through cuttings and tunnels too costly to contemplate, even with modern methods of construction and excavation.

Under the final mode of working by the Birmingham and Gloucester Railway, each descending passenger train halted at the summit and whistled for the brakesman, with three further whistles or short-sharp blasts to signify the checking and application of brakes. The Head Brakesman took command on the middle coach, the loco fireman stood by the tender brake while under-brakesmen and travelling porters—also the train guard—occupied positions assigned to them by the Head Brakesman. They communicated with each other by means of whistle and hand-signal codes. In the early days brakesmen were on continuous duty at Blackwell, except on Sundays, between 6am and 10pm. Sunday duties were supervised by the guard on each descending train. The B. & G. Company eventually fixed a vertical post as a marker, at the summit in Blackwell, surmounted by a lamp, that would be lit at night or in times of fog and mist. In fog and snow, banking engines for up services had to be buffered-up to their trains in Bromsgrove under the direction of a competent person on foot, walking at the side of the track.

In the days of steam working a large number of expresses and long distance passenger trains stopped at Bromsgrove, a tradition that continued

Afternoon delight. A solitary 'Jinty' 3F 0-6-0 number 47302 banks an unidentified northbound passenger express away from Bromsgrove on 22nd September 1956. The young man in the rear coach looks inquisitively towards the locomotive, wondering perhaps if one banker is going to be sufficient.
(Michael Mensing)

Modern Motive Power — class 45 diesel electric locomotive 45013 passes non-stop through Bromsgrove 'rebuilt' station with the 11.45am Cardiff-Newcastle service on 28th October 1978. Station buildings (left) await demolition. *(Michael Mensing)*

Ivatt class 4 2-6-0 number 43017 enters Bromsgrove Station with a local stopping train on 20th September 1956. The station was still gaslit even in this late era. McConnell's house, now demolished, is seen in the background. *(J. A. G. Coltas)*

47 540 races down Lickey bank on 19th August 1986 with the 09.36 Liverpool Lime Street-Penzance. It is seen here at the bottom, on the non-platformed downside of Bromsgrove Station. The rear of the train is just about to clear the crossover, which allows southbound services to use the one bi-directional platform. *(Stephen Widdowson)*

into the 1960s and the early part of the diesel era. Today most trains flash through at high speed, apart from a few multiple units forming the local Birmingham/Barnt Green-Worcester service. Whether north or southbound, these, and any other services, have to work into and from the former up platform, as the down platform—damaged in the 1967 derailment episode—was closed and demolished. As part of the rebuilding programme the remaining platform was reduced to a mere 200ft. All southbound trains calling at Bromsgrove, after descending the Incline, cross to, then occupy, the up platform line. On leaving they regain their correct running line via cross-over points to the south of the station.

Realignment of tracks, at the time of the station rebuilding in 1967, allowed speeds to be increased and the former 30mph speed limit abolished. In fact, during the 1970s the allowable speed was raised to the 70mph level.

Block trains for the Bromsgrove oil depot usually approach from the up direction, using the same crossover as passenger stopping trains. There are run-round facilities for the locomotive

Class 31s Nos. 31292 and 31178 are seen departing from the oil terminal at Bromsgrove on 13th November 1980 with empties bound for Port Clarence. *(C. J. Tuffs)*

within the terminal, which sometimes waits for the tank wagons to be unloaded but usually returns light engine to Birmingham. There is also a special siding reserved for crippled tankers.

Northbound freight trains requiring assistance wait at the south western approaches to the station and are usually dealt with immediately. If there is a following passenger train in close proximity the freight is drawn up on a loop and the faster service allowed to overtake. High Speed Trains needing assistance, perhaps through mechanical faults, cannot be banked from the rear as they lack normal buffering and coupling gear. Use has to be made of a drawbar concealed under the fore-end, designed for towing purposes only.

In the days of steam most trains were expected to need bankers, although a few light passenger trains, perhaps more than half empty, would be able to make it alone, especially if something happened to the banker. Those trains not calling at Bromsgrove Station halted at the southern approaches. Bankers made rearward contact but were not coupled, eliminating the need to uncouple at the summit. Signals were made between train engine and banker by shrill blasts of the whistle, known as 'crows'. An all-clear signal, allowing the train to start, was a long blast followed by two short double blasts recorded as x-xx-xx. This was answered by a single crow from the banker. The locomotives (in some cases three or more) would then open their regulators, trying to gain as much momentum as possible before feeling the rise of the gradient.

A continuous slog followed until reaching the summit, with the hardest work often left to the bankers, indicated, on many freight trains, by slack couplings. Leaving more work to the bankers was often necessary with steam-hauled passenger trains, helping to keep the fires intact for their later effort—beyond the summit—in the name of good time-keeping. It was impossible to judge the amount of effort expended by the train engine on passenger services on account of the close coupling. At Blackwell the bankers then, as now, dropped back, running into a special siding where they might await a chance to return. With three or more bankers these would drop off one at a time, only the first remaining in contact with the train into Blackwell Station. Several bankers would descend together to avoid congestion on the down line.

The stop for compulsory brake testing at Blackwell, in the down direction, appears to have been abolished in November 1943. All light engines, however, were forced to halt at the top of the Incline and returning bankers brought together un-coupled. In the latter case the lead engine going down would apply the hand brake at the summit, propulsion being from the rearward loco, or back-marker. Brakes on all engines were constantly adjusted to keep them together until reaching the level at Bromsgrove. Fully fitted freights and passenger trains were, at that time, restricted to 10mph through Blackwell Station, and not allowed to descend in under five minutes.

Due to so many vehicles working down hill with brakes applied, ballast at the side of the tracks, especially on the down side, was usually covered with a reddish-brown sprinkling of iron ore rust particles. Even with all brakes applied and spragged, in early steam days, some trains were known to have run out of control, passing through Bromsgrove

Western Region allocated a varied collection of their Hawksworth-designed 0-6-0 pannier tanks. Here, after buffering up, 8487 and 9493 start to assist a Black 5-hauled express out of Bromsgrove on 5th August 1961. *(John Edgington)*

'Jinty' tanks 47301/47303/47565 bank a heavy freight up towards Blackwell Summit on 11th July 1956. The engine drivers, along with the guard, seem relaxed as the trio of tanks lay a smoke pall over the surrounding countryside. *(John Edgington)*

Blackwell — 31st August 1955. 'Jinties' 47308/47305 prepare to 'drop off' through the station after banking a northbound freight.

(John Edgington)

full tilt, their whistles screaming, unable to stop for at least a mile towards Stoke Works. Fibres and flakes of red hot metal frequently spewed from the wagon tyres and, on dark nights, whole trains would seem to have caught fire but, in fact, were not in any actual danger.

The coaling and watering points for steam bankers were originally in the locomotive depot to the rear of the down platform at Bromsgrove, but later changed to the opposite side of the tracks on the southern approaches. Its coal road or siding was on a parallel but higher level, reached by an inclined plane from the station end. This was opposite a signal box from which instructions could be given concerning the number of bankers required for any given train. Banking engines normally took their duties in strict order of rotation, and only the minimum time was allowed for refuelling, watering and maintenance checks, important as these might have been. Water for tanks and tenders was drawn from a partly natural but enlarged and improved reservoir, known as Pikes Pool, at the side of the Incline towards the summit. It was engineered by the Birmingham and Gloucester Company, and proved a great saving in water rates!

In present day conditions the fastest trains can easily descend the Incline at the 'legal' limit of 80mph, there being no varying speed limits between Blackwell and Bromsgrove. However, loco-hauled up trains achieving the top mph limit at Bromsgrove cannot sustain this against the grade and the 75mph restriction at the summit is largely academic. High Speed Trains are by far the best performers going up. With 4,500hp available from both front and back a bottom speed of 80mph is usually only reduced to around 70mph by the summit. For heavy freight trains in the down direction, the limit is 30mph. A special form now allows an H.S.T. with one failed power unit to 'have the road' from Gloucester to climb the Lickey. It is issued from Gloucester Signal Box.

Looking back, very few of even the more powerful steam hauled trains could pass the summit at more than 25mph, and the normal banking pace for them may well have been below 20mph. In later years steam trains working down the bank were limited to 30mph, although freight, especially the unfitted type, was much slower, eventually limited to 11mph. All freight had to be halted and checked at Blackwell, before restrictions were lifted, on a special siding reserved for this purpose. With unfitted freights the guard had to make sure that at least every third wagon had its hand brakes pinned down. The releasing of these hand brakes also had to be supervised in a loop at Bromsgrove Station, where there was a siding for crippled stock—either damaged on the Incline or deemed unsuitable for the ascent. The final ruling concerning hand brakes on descending goods stock was that all brakes should be pinned at the summit, with one wagon brake pinned down for every two loaded vehicles, and one for every three empty vehicles.

A 'stable' of Lickey Bankers, seen on 24th February 1962, await their next call at the coaling stage, Bromsgrove. Note fresh supplies of coal awaiting unloading on the higher rail level. *(Michael Mensing)*

An afternoon up local service, on Easter Saturday 17th April 1965, approaches Bromsgrove Station. D6941 is on the stabling point access neck road whilst D1712 heads a down freight. *(Michael Mensing)*

THEN AND NOW

Class 31/1 number 31248, in charge of the 10.30 York to Exeter St. Davids relief service, is seen passing a simplified Bromsgrove on 8th August 1985. *(C. J. Tuffs)*

The 13.40 Paignton to Newcastle service is seen levelling out at Blackwell, at the summit of the Lickey Incline, on 8th August 1985 with power cars 43175 and 43034 providing the traction. High Speed Trains are by far the best performers up the Incline, and attain speeds not far below the official speed limit. *(C. J. Tuffs)*

...ALL BRAKES MUST BE PINNED DOWN.... A very important message to drivers of freight trains descending Lickey. This picture, taken in 1965 shows the 'brakesman', who was permanently employed by B.R. until all trains became fully fitted. The actual post was dispensed with in 1981 when Mr Michael Brean performed the last duties. The wooden stave was used to hold down the brake lever arm on every third or fourth wagon, depending on the train's length, to enable the brakesman to insert a pin into the appropriate hole located on the slide bracket. Each pin was mounted on a length of chain that hung from each slide bracket. Occasionally, although strictly against the rules, the brakesman could be seen riding upon the brake lever arm! *(W. H. A. Thompson)*

At Bromsgrove non-stop passenger trains or locomotives running light engine, on the centre road, were first limited to 10mph southbound or even forced to a halt until the 1890s. Speed on the Incline, for such traffic, was first 27 but later 30mph throughout.

Loaded traffic over the Incline was usually brisk in both directions but tended to favour down workings, as there were large coal trains running south from collieries in both the east and west midlands, but not compensated by return tonnage. When the West Midlands were more prosperous, and before the decline of freight traffic, nearly half the products and exports of this area worked down the Lickey Incline at some stage or other, even during the period of post-war reconstruction. There was always a fair cross-section of goods rolling stock to be seen, with mineral and hopper wagons of all types, pallet vans, oil tankers, cattle wagons, car transporters and, in more recent years, a whole range of freightliner and container traffic. In 1965 alone, 90,000 tons of freight were taken over the Incline in a single week.

In the days of steam medium to heavy freight needed two, three or even four bankers. Any goods train of more than eight loaded mineral wagons, twelve loaded goods wagons or sixteen empties had to be banked to Blackwell, which in practice usually meant all services.

On most days there would be at least six steam bankers on duty. Although the majority were tank engines, the first bankers or assistants were tender engines and a number of this type were reintroduced, at odd periods, throughout the years. No experts have yet enlightened the public as to the significance of maintaining one large engine and a number of smaller ones for banking, as frequently happened after the First World War. On Sundays only one large engine was needed, but it is difficult to understand why two large bankers were never ordered for weekday duties as a regular arrangement. Driving and firing a banker could be a demanding yet monotonous task, which their crews faced with the fortitude and endurance expected in British railwaymen of all generations.

It has been recorded that in April 1958 a short and fairly light passenger train, headed by a 'Standard' Midland Compound, was not receiving the assistance it had the right to expect from its banker (an ex-G.W.R. pannier tank). Deciding to work out its own salvation it pulled ahead at 35mph and left the tank engine far behind, unable to stay in the running. There have been several stories in this category but most seem apocraphal and without much basis of evidence.

The last recorded steam banking duty was undertaken by an ex-G.W.R. locomotive No. 6947 'Helmington Hall' on 7th October 1966. Number 6947 happened to be handy when a type 3 diesel-electric locomotive derailed at Bromsgrove and a temporary replacement was needed. The 'Hall' was normally based at Oxford and was of a class of locomotive seldom worked over the Incline.

The last regular steam banker shedded at Bromsgrove was the '9F' No. 92079, which inherited its massive electric headlamp from 'Big Bertha'. After leaving Bromsgrove it spent a further twelve months at Birkenhead, working iron-ore trains from the docks to Shotton Steel Works before being withdrawn for scrap.

Freight Then. Stanier-designed 2-8-0 number 48700 climbs the 1 in 37.7 on a hot August afternoon in 1960, banked by 9F class 2-10-0 number 92079. Note the guard's van next to the train engine along with the mixed collection of goods vehicles. *(John Edgington)*

Freight Now. Class 31/1 number 31303 is seen heading a northbound mixed freight up the Lickey at Vigo on the 18th November 1981 and is assisted in the rear by class 37s numbers 37254 and 37265 *(C. J. Tuffs)*

2-10-0 92079 at Derby Works in the late 1950s having been fitted with its headlamp before returning to the home shed, 21C Bromsgrove.
(Real Photographs/D.Sharpe Collection)

92079 at Bromsgrove South in 1958 awaits its next banking turn in the summer sunshine.
(A. W. Flowers)

Locomotives on and over the Incline

A number of typical locomotives, working on or over the Lickey Incline, have been recorded in other chapters, mainly in chronological order. It would be tedious to repeat every detail and dimension of these but a few further notes may be of interest.

For nearly a century and a half a wide variety of steam engines, familiar over other parts of the national system, spent at least some time on the bank. This was especially true of tank engines for the banking duties. At one time it seemed almost a matter of course for a few representatives of each class to put in an appearance at Bromsgrove Shed. These would have been designed by Kirtley, Johnson, Deeley, Fowler and other important locomotive engineers.

Exceptions for banking duties but seen on the Incline in normal or special traffic, included the 'Standard' Midland Compounds. After the 1923 grouping a number of these locomotives were brought south of Birmingham but still seemed to have been excused banking duties. Class '4' 0-6-0 goods engines and the '999' class of 4-4-0s were rarely seen in the area. Some of the latter, however, may well have passed over the Incline when transferred to the Somerset and Dorset line for a trial period in 1924/25.

Before the grouping one of the most attractive express engines to work between Bristol and the north east were Johnson 4-4-0s of the '700' class, mainly based at Derby. They were often partnered or assisted by a group of rebuilt 'Class 2Ps' from Bristol.

Apart from actual bankers there were very few tank engines working trains over the Incline, at least until some time after the grouping. In comparative terms there were very few local or stopping trains, although most long distance trains stopped at Bromsgrove, heading north or south. Most locals from Birmingham New Street, hauled by tank engines, seemed to disappear onto the Redditch line before the top of the Incline, diverging at Barnt Green after working through the Birmingham suburbs.

Typical representatives of the pre-1914 era would be Johnson 2-4-0s and 'Class 2P' 4-4-0s, double-framed Kirtley tender engines (0-6-0s) and a number of Johnson single-wheel 'Spinners'. Tank engines, other than bankers, mainly ran light engine through Bromsgrove, either on trials or between Bristol and the Derby Works for repairs. Soon after the grouping there was a plethora of

Busy time at Blackwell, 7th July 1956. 2P 4-4-0 number 40426 departs from the down platform with a Birmingham-Bristol Temple Meads local service. At 4.45pm the up platform is crowded with mums and dads along with children of all ages. This picture epitomises Blackwell in those steamy days of the 1950s! *(Gavin Morrison)*

Gloucester-Birmingham stopping train banked by a solitary 'Jinty' on 7th July 1956. The trains' vehicles are a mixed bag, the leading coach being L.M.S. wooden suburban stock. *(Gavin Morrison)*

Horwich types inherited from the Lancashire and Yorkshire Railway Company (taken over by the L.N.W.R. shortly before the grouping), especially 2-6-0s ('Crabs').

Stanier-designed engines came a few years later, in all shapes and sizes supported by the 'Standard' Midland Compounds mentioned before. 'Royal Scots' by Fowler and even a few Pacific types made rare rather than regular appearances, while there were occasional visits from 'foreign engines', either in wartime or hauling specials. The 'Jubilee' was the mainstay of fast trains and passenger express workings, from the mid-1930s to the era of dieselisation. 'Black Five' Staniers were among prevalent examples during the 1950s, a few appearing with Caprotti valve gear and roller bearings, others having double chimneys. The 'Jubilees', however, as an improved development of the 'Patriots', were perhaps the most stylish performers,

Saturday Special! Down Gloucester relief pictured half a mile down Lickey on Easter Saturday 20th April 1957, headed by Stanier-designed class 5 4-6-0 number 44810. *(Michael Mensing)*

Those were the days! — Jubilee class 4-6-0 number 45597 'Barbados' gets to grips with a Bristol-York express during one of the fine summer days of 1956. Note the Fowler style tender, less common on Jubilees than the larger — and more stylish — Stanier type.
(Roger Carpenter Collection)

What a relief! — an unusual locomotive rostered to work the down Bradford-Bristol relief express on Saturday 13th September 1958. 42791 was built initially for freight duties by the L.M.S. These engines could, however, if called upon, produce some lively passenger turns. Pictured near Vigo.
(B. S. Moone)

whilst a number of 'Patriots' were seen on all types of traffic. Even after the Second World War 'Standard' Compounds were still fairly numerous, as also 'Class 2Ps' until the late 1950s.

The 'Jinties', Panniers and other 0-6-0T types, although less glamorous than the larger tender engines, were some of the toughest and hardest working locomotives in the land, to say nothing of their performance as bankers over the Incline. In British Railways days each banker, large or small, was in steam for a fortnight at a stretch. At the end of this time they returned to the running shed for a check-up and boiler washout. (General overhauls were usually every twelve months). They were normally serviced in the open on the banking spur, ready to move off either on to the main line or goods line to the rear of any train needing assistance, at a moment's notice. Returning light engine from Blackwell they crossed from the main southbound line to their small yard beyond Bromsgrove Station.

The first duty of an incoming crew (often with triple manning), at the start of their shift, would be to clean the fire and make an all-round general inspection. Although still in the line of duty, this was the only respite in a normally tough stint.

Although sometimes working as triple units (less frequently in fours), the equivalent of the single ex-L.N.E.R. Garratt locomotive, most of the 0-6-0s worked in tandem, equal to 'Big Bertha'. This was the a general requirement for the majority of workings.

Tonnage of trains approaching Bromsgrove in a northerly direction was telegraphed from Gloucester, Cheltenham or Worcester. As the train in question passed through from Stoke Junction (Stoke Works), the Bromsgrove signalman displayed to the waiting banker crews a large numbered tablet, corresponding with the number of bankers needed. Exactly one minute was allowed for them to make contact with the rear end of their train. Approaching train engines also signalled their needs to the boxman at Stoke Junction with one, two or three whistle crows.

Diesel train power was mainly drawn from the 45 and 46 classes although 47s became popular at a slightly later period. A number of types and classes from other regions or companies were sometimes forbidden on account of certain restrictions and clearance difficulties.

A quartet of bankers. A busy time for the Bromsgrove-based 'Jinty' 0-6-0s as they render assistance to an up express at Vigo in the mid-1950s. The rear locomotive appears to be 'doing all the work'. *(John Edgington)*

BANKERS OLD AND NEW

Class 45/1 number 45102 is seen passing through Bromsgrove on the 15th September 1982 in charge of the 6.00hrs Aberdeen to Bristol Temple Meads. The class 37s in the sidings on banking duties are 37231 and 37294. *(C. J. Tuffs)*

A Footplate Tale

Based on the memories of Derek Barton of Bromsgrove, whose early career years were spent in 'working on the bank'

Derek began his working life in 1945, shortly after leaving school at the age of fourteen. Although employed as an errand boy for the Co-operative Butchery Department, the pay was far from satisfactory but extra cash could be earned by working for the railways over the weekends.

He was employed by the L.M.S. and, after nationalisation, by the London Midland Region of British Railways. He spent most of his Sundays in the loco sheds at Bromsgrove m.p.d. and later on the coal road of the banking spur. He was a member of a gang of youths, all fifteen years old and over, allowed to engage in weekend stints to improve their earnings and assist the railways at a time of acute labour shortages. The work evolved around locomotives of all types, shovelling coal out of wagons into tenders and bunkers, later extracting rubbish from the ash pits for disposal in other wagons. 'Big Bertha' and the 'Jinties' all needed re-stocking with coal after numerous various journeys up and down the Incline, assisting passenger and goods trains.

Derek relates *'You had to be sixteen years of age before the railways could employ you on a full-time basis. As soon as my 16th birthday came I said goodbye to the errand bike at the Co-op. It was customary to be taken on as a cleaner on the loco side. I never actually cleaned an engine; my first week at Bromsgrove, in a full-time capacity, was spent doing the same jobs I had been doing on many previous Sundays. The second week was taken up with tuition in basic firing techniques, at Saltley Shed, Birmingham. After being passed-out I was made a 'Registered Fireman'.*

'The first duty, I remember, was on a 'Jinty' banker, No. 7301. However, I did not have long to wait before my turn came to fire 'Big Bertha', then numbered 2290. Most of the drivers in those days were pleased to help and advise you, provided you were willing to listen. 'Big Bertha' was a lovely engine to work on, whether banking express or goods trains up the Incline. I always marvelled at the impression of smooth power this locomotive gave. One particular driver was a pleasure to fire for. He had done it all, over the years, always able to instruct you or demonstrate how things could be done, whether it was a driver's job or a fireman's duty. He was, without doubt, a good engineman, able to do the job well without wasting coal or excessive use of steam, as some footplatemen might have done. Yet one must not decry the other drivers, as most of them were a pleasure to work with, although a few may not have liked me.'

'Sometimes we had inspectors on the engines for a 'trip up the bank', to assess our capabilities in their eyes. I can recall, one hot summer's day, an Inspector Wood climbing into the cab of our banker, just as we were about to assist a local passenger train out of Bromsgrove Station. I was teamed-up with a driver named Vic for this particular duty. Now Vic was a gentleman (in the most important meaning of the word), always willing to help. Before we came up to the platform he assisted me in 'lumping the engine' (large lumps of coal put into the firebox to save the shovel and keep the dust down on a warm day). I thought it better to use the shovel once we were on the bank. As I swept up the footplate and was about to shovel the dust into the firebox, Mr Wood joked 'Enough in there, lad' as though prompting me in an elementary duty. Vic smiled. On returning down the bank I sat on my side, while Vic sat on his with Mr Wood in the middle of the footplate. As we were now running tender first coal dust was soon swirling round the cab, a strong breeze activating it even more, and—by the time we reached the foot of the bank—the inspector looked more like Al Jolson than a professional railwayman!'

'Another memory of Mr Wood was...'later in my railway life, when firing 'Big Bertha' at the rear of a very heavy freight train. The inspector was his usual pleasant self, always ready to ask and answer questions... Mr Wood remarked about 'Bertha', then steaming well... 'Derek, this engine is about to blow up, what steps are you going to take?' I replied, without hesitation, 'Bloody big ones—coming?'

'The most talked about engine that came to Bromsgrove, in later years, was the Garratt, No. 69999. Now this was something...a mighty engine but no blast from the chimney—just a dull roar. Coal came in at one end and out at the other. We did not need a shovel on that locomotive—a wheelbarrow and plank would have been more appropriate. It is said that the men at Bromsgrove did not like No. 69999. In my view the Garratt was not built for use on the Lickey Incline. Why the London Midland Region or the L.M.S. never built another banker like 'Big Bertha' I'll never know.'

'Yes, those days at Bromsgrove were great times, with many memories. Harsh, cold winters and hot summer days were all part of the life. It certainly gave one an appetite, not the least of my enjoyments being the railway's method of frying eggs, bacon, sausage and fried bread on a clean shovel—it was all part of the life. After eight years I was again tempted by money, when a job came along with no shift-work, no Sunday duties and more cash in the pay packet. I left the Incline and the railways for good but not without mixed feelings and some regrets.'

A few memories of the Incline from John F. Smith C.Eng. M.I. Mech Eng. F.I.E.D., conclude this final section...*'In 1931 I experienced a very exciting*

Routine work for the drivers and firemen of Bromsgrove as 58100, with an unidentified 'Jinty' tank, bank an up freight near Vigo circa 1949/50. *(John Edgington)*

episode on the Lickey Incline, while I was still a youth. It was a very hot summer afternoon, about three o'clock. I was at Vigo watching the trains. A heavy freight was hammering away at 'the mighty conflict', the wagons going slowly past me. Suddenly I saw there was a wagon on fire. It was a wooden-sided type and the flames were leaping 6ft. high—the side planks were well alight. At first I was too astonished to take it in. It seemed like a load of fire, being transported somewhere on schedule.

'The brakesman obviously could not see it, because high-sided vans hid it from view. It was also hidden from the footplate crew for the same reason. As the brakevan came level with me, I was dancing up and down and shouting...'Your train's on fire...Your train's on FIRE!' The brakesman looked up at me as much as to say 'Cheeky young devil...you're just having me on!' However, I must have been convincing and made some impression. Perhaps he could smell the burning, if the wind changed in his direction. After a minute or so he stopped the train on mid-incline and applied the brakes to the rear half, uncoupling the blazing wagon which was taken to Blackwell where the brigade was summoned. I raced round to Blackwell to witness the final quenching of the flames, along with the re-coupling of the rear part of the train. As I understood it, extra bankers were called up from Bromsgrove to give the train a final boost to the top. The wagon contents were a load of tiles packed in straw, ignited by a spark from the train engine.'

'I can remember, in 1934, when the L.M.S. Garratt came to help out on the Lickey Incline for a brief period (more in the nature of tests). As a boy I spent some time train watching at Vigo Stile, about half-way up the bank, along with other children. We never knew, until the locomotive was upon us, if it was the Garratt banking or the decapod 'Big Bertha'. The 0-10-0 we knew—perhaps too well—but if it was the Garratt, there was a shout of triumph...it was so exciting to see this huge double-engine pounding up behind a heavy freight train. It was, however, soon withdrawn—partly because of its massive fuel consumption, requiring a ton of coal to take it up the Incline.'

'Only a few weeks ago (at the time of writing)—after a lifetime of railway watching on the Lickey Incline, I experienced another interesting event. On Wednesday 17th January 1988 I was close to Bromsgrove Station and could hear the throb of diesel engines. I walked on to the platform and there was a short civil engineering train, standing in the station but heading towards Gloucester, positioned on the 'up' line waiting to be crossed over to the 'down' line. The locomotive, a class 37 diesel-electric, was coupled to a 'Matisa' ballast cleaning machine—a huge piece of rolling stock painted yellow for safety—twice the length of the locomotive. This machine would be capable of excavating ballast to the depth of two feet, washing, cleaning and returning it to the track, with dross deposited into another wagon. Coupled to the machine were two closed vans—wooden-sided and about 40 years old. Two long-wheeelbased, wooden-sided wagons from about the same era, made up the rest of the train.'

'I strolled along for a few words with the young engine driver. He said the ballast cleaning machine was being taken back to Gloucester Area headquarters. I asked him how many miles his class 37 had covered. He replied 'over a million miles'. He expected it would be still running into the next century. How amazing!'

Driver's eye view! Pictured from the footplate of 58100 on 26th May 1951. 'Big Bertha' is about to bank a heavy freight over Lickey, leaving Bromsgrove Station. *(H.C. Casserley)*

Class 47/3 number 47307 prepares to leave southbound from Bromsgrove platform with empty oil wagons to Milford Haven on the 9th December 1980. It was here that Derek Barton talked to the driver of the class 37. *(C. J. Tuffs)*

'The Devonian' and 'The Pines Express'

Bournemouth bound! 'The Pines Express' descending Lickey, with ex-L.M.S. 4P 4-4-0 3-cylinder compound number 41049 piloting a Stanier-designed L.M.S. 'Black 5' 4-6-0 during the mid 1950s. *(Michael Mensing)*

'The Devonian' was perhaps the most outstanding of the regular express workings over the Lickey Incline. This travelled between Bradford and Paignton, first using its official name in 1927. Shortly before the Second World War, in the summer of 1939, it made the complete journey of 323 miles in 8 hours 26 minutes southbound and 8 hours 29 minutes northbound, the return journey being slightly longer on account of the delay at Bromsgrove and timings over the Incline. By 1953 the schedules had deteriorated rather than improved. Leaving Bradford at 9.22am it arrived in Paignton at 7.20pm (8.6pm on Saturdays), with a southerly running time of 9 hours 47 minutes and an up or return journey of 9 hours 58 minutes. There was an improvement towards the end of steam workings but, even then, not up to later pre-war standards. Leaving Bradford at 10.15am (fifteen minutes earlier than in 1939) it arrived in Paignton (1959/60 at 7.15pm in exactly 9 hours, returning to Bradford after a run of 9 hours 33 minutes. Both up and down services passed through Bromsgrove Station within a quarter of an hour of each other. Motive power was usually a 'Jubilee' 4-6-0 throughout.

It was difficult to explain why the running times in later years deteriorated, when most people would have expected either improvements or a return to former standards. Perhaps it was partly the result of a deliberate policy concerning motive power. Targets had been set for higher monthly mileages from each engine, which resulted in difficult and complex rosters. This, in turn, led to neglect of servicing, so that frequently engines were over-worked with insufficient maintenance to meet their diagrams. Later running and more relaxed overall schedules were the obvious outcome of neglect.

By 1964 there was a speed-up on the north-east/south-west route to 8 hours 17 minutes southbound and 8 hours 16 minutes northbound, improving the pre-war timings by a narrow margin. With the full dieseliation of the line an even greater improvement began to emerge. In the 1974/75 timetable workings had been cut to 7 hours 7 minutes (down) and 7 hours 11 minutes (up). On Summer Saturdays the northbound service was frequently cut to 6 hours 55 minutes, including a scheduled stop at Gloucester, although banking services were no longer required at Bromsgrove. Unfortunately the 'The Devonian', as a named express, was withdrawn shortly after these improvements had been made.

The appearance of the High Speed Trains has not greatly altered or improved the north-east/south-west services. They mainly fit into earlier schedules as planned for non-HST (diesel) workings, but are capable of greater effort and performance than at present allowed. At an average speed of 100mph it should be at least possible in the near future to make a connection between New Street (Birmingham) and Bristol (Temple Meads) in a little over an hour.

'The Pines Express', like 'The Devonian', commenced running under its popular name and style, from 1927. It originally ran from Manchester to Bournemouth via Birmingham (New Street), Bromsgrove, Bath and the Somerset and Dorset line. According to the summer timetable for 1953 it left Manchester (London Road) at 10.25am, reaching Bournemouth West at 5.35pm. The Saturday run commenced at 10.25am, arriving at 6.8pm. The northbound journey departed from Bournemouth West at 9.45am, reaching Manchester at 4.36pm. The Saturday service, leaving at 9.25am, reached Manchester (London Road) at 4.48pm. The winter timetable for 1959/60 gave the Manchester starting time as 10.15am, reaching Bournemouth West at 5.32pm, which was 7 hours 17 minutes—an hour longer than similar trains scheduled in the era before the First World War. The return journey was 9.45am to 4.52pm or 7 hours and 7 minutes. There were 40 minutes between southbound and northbound services passing through Bromsgrove and over the Incline. The usual motive power, especially between Birmingham and Bath, would be a Stanier 'Black 5'. From 1962, shortly before the closure of the Somerset and Dorset line, the 'The Pines Express' was transferred to the Western Region on a route via Snow Hill (Birmingham) and Oxford, improving journey times to 6 hours 44 minutes.

By 1962 the 'The Cornishman', previously a named train of the Great Western Railway and later the Western Region of B.R., was transferred to run via the London Midland Region. It formerly ran from Wolverhampton to Penzance, changing to run via Birmingham New Street rather than Snow Hill in September 1962 (the winter timetable change). During 1964/65 it departed from Sheffield at 8.12am arriving in Penzance at 5.35pm. The northbound or 'up' express left Penzance at 10.45am, reaching Sheffield by 8.40pm.

'The Devonian' roars over the Lickey summit on a warm summer's day in 1959 with Jubilee class 4-6-0 number 45572 'Eire' in charge. A picture that would soon change with the demise of steam and semaphore signals, which had been part of everyday railway life for many years. *(B. S. Moone)*

More About 'Big Bertha'

This 0-10-0 four-cylindered giant was the only British decapod active at the time of building. Although designed before the First World War, she was not launched in active service until nearly two years after the War. One of her outstanding features was the cab-tender, with fully-enclosed protection, thought necessary as she spent half her running time in reverse. The sides of the tender, however, were reduced to allow improved vision when running down the Incline. She had two super-heated boilers, which were refitted alternately throughout her working life.

The first steam tests took place early in January 1920, so that she arrived at Bromsgrove Shed on the 20th of the same month. She remained at Bromsgrove for 36 years, with brief visits to Derby Works and other centres for repairs and overhauls. She was also given brief trials on coal trains between Toton and Brent in 1924, shortly after the grouping, but this was not a success. For a short period in 1949, while returning south after attention at Derby, she assisted in banking operations at Camp Hill.

The distinctive headlamp was not fitted until 1921, and was useful for buffering-up after dark or in misty conditions. It was powered by a steam-turbo generator, the latter fitted below the running plate near the front left-hand side of the cab.

Other, perhaps more important items, among her distinctive features were ten coupled wheels, four (simple) expansion cylinders and the system of steam distribution. In the latter case two ordinary piston valves regulated steam supply to all four cylinders.

Up to that period the only decapod on a British system was a tank engine of the Great Eastern Railway Company, making 'Big Bertha' the first British tender engine with five coupled wheels per side. A version of Walschaerts valve gear was used, arranged with a cut-off of 75 per cent, driving an external admission piston-valve above the outside cylinder. It was a unique example, at least in Britain, of a four-cylindered engine of the heavy freight type worked by simple expansion. No rocking levers or extra valve gear, above what was required for a two-cylinder locomotive, were needed. It thus avoided the complications for repair and maintenance normally involved with multi-cylindering.

Drive from the cylinders was direct to the third or centre wheels, allowing for the placement of suitable connecting rods for all sets of motion, with four cranks fixed upon the central axle. The second axle was slightly cranked in two places, allowing connecting rods to have satisfactory clearance, also to avoid unduly raising the inclined cylinders. Reversing was by means of a servo-motor (steam operated) near the reversing arm shaft, fed by pipes from the footplate—controlling the cataract valves.

Suspension of all wheels was in the form of steel laminated leaf springs—with compensation beams or bearings between first, second, fourth and fifth axles. Steam brakes, on the first three pairs of wheels, were actuated by a cylinder behind the crank shaft, while the two rear pairs depended on an independent cylinder below the drag box. Both footplate and tender hand brakes were also used in a reverse capacity.

The boilers were furnished with Midland Railway type super-heaters, each having 27 elements. The firebox was of 10ft. length having a grate area of 31½sq.ft. The boilers were tapered in two directions, the crowns sloping in a downwards direction towards the backplate. This latter was similar to the backplate featured on the Midland Standard Compound. Designed for a working pressure of 200lbs. per sq.in., the actual pressure at which the safety valves discharged was 180lbs. p.s.i. Water feed was via two injectors, water previously warmed in an apparatus on the covered tender, using steam from the 3⅛in. safety valves, that would otherwise blow to waste.

Being within easy reach of fuel and water, the tender was of moderate size. Its main feature of interest was the neatly designed cab-shelter. Maximum weight on the centre axles was 15½tons.

Among a few other misfortunates, 'Big Bertha' was rammed and badly damaged, at the front-end, by a 'Jinty' (No. 47305) in November 1953. This required attention at the Derby shops. Also during the same year there were slight alterations to her front end appearance and to the main frames, made during the course of major overhaul. This latter entailed the welding of a new section of 1in. thickness extending between the leading driving wheels and the buffer beams.

In December 1953, while banking near the summit at Blackwell, a connecting rod failed, injuring the cylinder, which had to be cut away before the engine could be moved. She was later towed to Derby, where a new cylinder was fitted.

Under the control of the Western Region, target numbers were introduced for all bankers, carried on the left-hand front lamp bracket. 'Big Bertha' was always No. 1 at Bromsgrove, although while off-duty or away for repairs, her replacement engine carried the diagram No. 1 or 1A.

By 1955 the first of 'Big Bertha's' super-heated boilers was considered unsuitable. The second boiler was then fitted, as a replacement, but also became suspect, leading to the withdrawal of the locomotive from service.

'Big Bertha' being fitted with its headlight at Derby works in 1921. The powerful generator for supplying the current to the lamp itself is clearly visible. The temporary flimsy-looking wooden scaffold certainly would not be tolerated under today's Health and Safety at Work Act. The sighting poles or bars are an addition fitted after recommendation of the loco crews — no doubt it was difficult to judge distance to the rear of a train to be banked, particularly at night.
(National Railway Museum)

All lit up! Another study of 'Big Bertha' outside Derby works in the winter of 1921 during alignment trials. Note the way the lamp beam diffuses to give downward as well as forward illumination. The guards van is the guinea pig for the focusing adjustments. What a job it must have been setting that great headlamp on a cold and frosty night! *(National Railway Museum)*

Leading Dimensions of 'Big Bertha'

Cylinders (x 4) diameter:	16¾in.
Stroke of pistons:	28in.
Piston valves (x 2) (outside admission type):	10in.
Outside lap valve:	0⅞in.
Inside clearance valve:	0⅛in.
Wheels, (diameter of tread):	4ft.7¼in.
Coupled wheel base:	20ft.11in.
Boiler diameter (outside front ring):	5ft.3in.
Length of boiler barrel:	14ft.
Firebox width (outside):	4ft.0⅝in.
Firebox length:	10ft.
Length of tubes and flues:	14ft.4¾in.
Diameter of tubes:	1⅞in.
Diameter of flues:	5⅛in.
Number of tubes:	147
Number of flues:	27
Heating surface of firebox:	158.25sq.ft.
Heating surface of tubes:	1,560,00sq.ft
Heating surface (total):	1,718,25sq.ft.
Superheating surface:	445,00sq.ft.
Grate area:	31,00sq.ft.
Boiler Pressure (designed):	200lb. per sq.in.
Boiler Pressure (working):	180lbs. per sq.in.
Tractive effort (at 80% boiler pressure):	40,7765lbs. (18.2 tons)
Adhesive power (without sand):	33,148lbs. (14.8 tons)
Adhesive power (with sand):	44,713lbs. (19.9 tons)
Total weight of engine in full working order:	73 tons 13cwt. 1qr.
Weight per foot run on wheelbase:	2.29 tons
Weight per foot run over buffers:	1.74 tons
Tender, tank capacity:	2,050 gallons
Tender, tocal capacity:	13ft.
Tender, total weight in working order:	31 tons 11cwt. 2qr.
Tender, wheelbase:	13ft.
Total weight of engine and tender in full working order:	105 tons 4cwt. 3qr.
Total length of engine and tender:	61ft.0⅝in.
Extreme length, above rail level:	13ft.4in.

58100 in its final form circa 1955, in British Railways black livery with simple lining plus the lion and wheel emblem. From the appearance, the paintwork has only recently been applied. The locomotive was kept reasonably smart right up to its withdrawal in 1956.

(8F Locomotive Preservation Group/Derek Harrison Collection)

Locomotives of the Birmingham and Gloucester Railway

No.	Name	Built	Builder	Wheel Arrangement	MR Number	Notes
1	Bromsgrove	1839	Forrester	2-2-2	Unknown	Two Forrester locos were
2	Tewkesbury	1839	Forrester	2-2-2	Unknown	absorbed into M.R. stock—
3	Worcester	1839	Forrester	2-2-2	Unknown	renumbered 243 (1) and 244
4	Cheltenham	1839	Forrester	2-2-2	Unknown	(1) but uncertainty as to their identity
5	England	1839	Norris	4-2-2		Sold privately 1842
6	Victoria	1839	Norris	4-2-2	277	
7	Atlantic	1839	Norris	4-2-0	-	Sold 1846
8	Colombia	1839	Norris	4-2-0	-	Sold to Aberdare Rly 1846
9	Birmingham	1840	Norris	4-2-0	-	Sold to W. Norswich 1846
10	Gloucester	1840	Norris	4-2-0	-	Sold to Taff Vale Rly 1845
11	W. S. Moorsom	1840	Norris	4-2-0	-	Sold to Taff Vale Rly 1845
12	Washington	1840	Norris	4-2-0	-	Sold 1846
13	Philadelphia	1840	Norris	4-2-0	271	Withdrawn 1856
14	Boston	1840	Norris	4-2-0	272	Withdrawn 1856
15	Baltimore	1840	Norris	4-2-0	281	
16	Moseley	1840	Norris	4-2-0	-	Sold December 1843
17	Pivot	1840	Norris	4-2-0	278	
18	Bredon	1840	B. Hick	4-2-0	279	
19	Defford	1840	B. Hick	4-2-0	280	
20	President	1840	Norris	4-2-0	282	Sold by Midland Rly 1852
21	William Gwynn	1840	Norris	4-2-0	273	Sold by Midland Rly 1852
24	Spetchley	1841	B. Hick	4-2-0	283	Sold March 1847
25	Eckington	1841	B. Hick	4-2-0	-	Sold October 1846
26	Ashchurch	1841	Nasmyth, Gaskell & Co.	4-2-0	-	Sold October 1846
27	Droitwich	1841	Nasmyth, Gaskell & Co.	4-2-0	284	
28	Pershore	1841	Nasmyth, Gaskell & Co.	4-2-0	-	Sold April 1846
29	Upton	1841	Nasmyth, Gaskell & Co.	4-2-0	285	Sold March 1847
30	Lifford	1841	Nasmyth, Gaskell & Co.	4-2-0	286	Sold October 1847
31	Niagara	1842	Norris	4-2-0	274	Sold by Midland Rly 1851
32	New York	1842	Norris	4-2-0	275	Sold by Midland Rly 1855

The following group were ordered by the Birmingham and Gloucester Railway, but paid for by the management of the Bristol and Birmingham Railway.

No.	Name	Built	Builder	Wheel Arrangement	M.R. Number	Notes
22	Stratford	1845	Sharp Brothers	2-2-2	53(2)	
23	Atlas	1845	Sharp Brothers	2-2-2	54(2)	
33	Evesham	1842	E. Bury	2-2-0?	)	Very little is known about
34	Kempsey	1842	E. Bury	2-2-0?	)	these two locomotives
35	Wadborough	1844	Sharp Brothers	0-4-2	214	
36	Bristol	1845	Jones & Potts	0-6-0	)	Goods engines supplied to
37	Hercules	1845	Jones & Potts	0-6-0	)	the Bristol & Birmingham Joint Committees. Acquired by the M.R. and renumbered 167 (1) and 168 (1) respectively. Both rebuilt in 1848
38	Great Britain	1845	Bromsgrove Works	0-6-0ST	276	Acquired by the M.R. and renumbered several times 276 (1), 105 (2), 300 (2) and 221 (3) rebuilt as a well tank in 1853

Note: This locomotive was taken out of service October 1861. Recorded as 'broken up' July 1862.

Rules for Working the Lickey Bank—Birmingham

Descending

1. At a point half a mile from the summit, there is a timber arched bridge, painted white. All engines must have their steam shut off here.
2. Near the top of the Plane there is a post on which a white light is kept burning. Trains must stop here.
3. Before stopping the driver will whistle for the Brakesmen, as well as giving three sharp whistles for the application of brakes by the Train Guard.
4. The Switchman is also to ring a bell as a signal to the Brakesman as soon as a train is in sight.
5. The train having stopped, the Brakesman will mount and take complete charge of the train.
6. The First Brakesman will mount the middle carriage. The Second Brakesman and others (Travelling Porters who may be on the train) will all take their orders from the First Brakesman, to distribute themselves — at least one to every three wagons of a wagon train. The Fireman will station himself at the tender brake.
7. It is hereby positively ordered that the Enginemen, Guards, Porters, etc, shall submit to, and implicitly obey the First Brakesman. They are to keep the brakes positively and firmly in hand ready to act.
8. No train shall descend without a Brakesman.
9. If the First Brakesman cannot accompany the train, the Second Brakesman will take charge.
10. The speed for a descending coach train shall never exceed 20 miles per hour in dry weather, or 16 miles per hour in slippery weather. The descending speed of wagon trains shall in no case exceed 10 miles per hour.
11. The First Brakesman shall make use of the following signals:

    One blast of the whistle: Start the train

    Two blasts of the whistle: Driver to shut off steam when the train has started.

    Three blasts of the whistle: Apply brakes.

    Note. This signal applies anywhere on the Plane and if repeated elsewhere than on the summit it means:- Apply brakes more forcibly, but gradually. One blast of the whistle (when on the Plane): Ease brakes gradually. Five blasts on the whistle means: Obstruction, Danger.

    Note. In daylight, warning of danger will be given by the showing of a red flag. At night, a red lamp will be shown.

Steam to spare! L.M.S.-built class 5 4-6-0 number 44688 breasts the summit at Blackwell in the late 1950s. All the period furniture abounds — Midland Railway signs, lamp posts, brakesman's hut, and 'All goods trains to stop to pin down brakes' warning sign. *(Tony Haynes)*

A summer Saturday at Vigo! A crowd of spotters and enthusiasts watch class 9F 2-10-0 number 92008 bank a military freight over the minor road bridge on the 17th August 1955. *(Michael Mensing)*

and Gloucester Railway Co (circa 1841)

Ascending

1. All trains stop at Bromsgrove
2. When an Up train is in sight, the policeman shall ring a bell.
3. The bell shall be a signal for the bank engineman to place his engine under the bridge at the foot of the Plane on the west line.
4. When the train stops, the Fireman of the bank engine will connect his engine to the train, which will then start when signalled by the Guard.
5. It is the special duty of the policeman at Bromsgrove to see that the carriages and wagons are properly connected both by centre and side chains, also that the last carriage or wagon has a brake attached.
6. The Guards are to stand to the brakes in ascending the Plane, keeping them constantly ready for use. A Guard or Porter must invariably ride on the last carriage or wagon.
7. As soon as the rail of a train reaches the summit of the Plane, the Fireman of the bank-ending will throw off his engine from the train, and at the same moment the train engine is to slacken speed by shutting off steam, and thus allow the bank-engine to go at least 30 yards ahead, and turn into the siding prepared for the purpose.
8. When the train has passed on, the bank-engine will descend the Plane on the east line at a speed not exceeding 12 miles per hour.
9. Any obstruction at the top of the Plane will be notified, in clear weather, by the showing of a red flag. At night or in fog, a bright red light will be shown.
10. It is the duty of the Switchmen both at the summit and at the foot of the Plane to keep the lines of way constantly clear.

ANY SERVANT OF THE COMPANY DISOBEYING THESE RULES WILL NOT ONLY MEET WITH INSTANT DISMISSAL AND LOSS OF ANY PAY DUE TO HIM, BUT WILL ALSO BE PROSECUTED AT LAW FOR NEGLECT OF DUTY

By order of the Directors

Willm Burgess
Gnl. Superintendent.

Special Instructions for Ascending

(An abstract from British Railways (Western Region)

Bromsgrove to Blackwell (Lickey Incline)

No train must ascend the incline without an assisting engine being immediately behind the last vehicle, except passenger trains formed of not more than four bogie wheels not exceeding 90 tons, and freight trains equal to not more than eight wagons of mineral, which can go up the incline unassisted by a bank engine, provided the last vehicle of a freight train is a brake van in which the guard is riding.

Up passenger trains composed of not more than 12 bogie vehicles and which are not booked to stop at Bromsgrove Station for traffic purposes, requiring the assistance of a bank engine, must be brought to a stand with the engine opposite the indicator (illuminated at night) fixed about 15 yards to the rear of Bromsgrove Station up home signals. Up passenger trains composed of more than 12 bogies and which are not booked to stop at Bromsgrove Station for traffic purposes must be brought to a stand with the rear vehicle clear of the connections at Bromsgrove South Signal Box, to enable the bank engine to get to the rear of the train.

In case of Up passenger trains which require to stop at Bromsgrove Station for traffic purposes and which require the assistance of a bank engine—it will be placed at the rear of the train while standing in the station.

In the case of trains other than passenger trains which require the assistance of a bank engine, it will be placed at the rear of the train at Bromsgrove South or Bromsgrove Station signal box as required.

When an up passenger train not booked to stop at Blackwell is required to stop specially at the station for traffic purposes, the driver of the bank engine assisting the train must be advised before leaving Bromsgrove that the train has to stop at Blackwell. The Station Master at Bromsgrove will be informed of such extra stops and he will be held responsible for advising the Driver of the bank engine.

Bank engines assisting trains up the incline must each carry a white head lamp and red tail lamp.

If a train fails on the incline and requires assistance, the Standard Rules and regulations must be carried out.

Bank power for up trains composed of coaching stock is provided in accordance with the tonnage wired to Bromsgrove Station from the last stopping place.

Drivers of trains composed of coaching stock requiring assistance in excess of that laid down for the tonnage conveyed should give the required whistle at Stoke Works Junction.

A telephone is provided 50 yards in advance of the Lickey signal to enable Trainmen to communicate with the Signalman at Blackwell.

The Up Line is worked under special arrangements. Catch points exist in the up line 300 yards in advance of the Lickey signal and near the top of the incline.

Up freight — Standard 9F 2-10-0 number 92155 draws forward at Bromsgrove South in readiness to be banked on 21st March 1964. *(Michael Mensing)*

and Descending Includes
Sectional Appendix to the Book of Rules, October 1960)

Lickey Signal.

A two aspect colour light stop signal, known as the Lickey signal, is fixed on the up side of the line 1,570 yards in advance of Bromsgrove Station box up starting signal and applies to the up line.

The Lickey signal must not be passed at Danger unless the Driver has received a ticket at Bromsgrove Station box informing him that the apparatus has failed, which will authorise him to proceed cautiously to the home signal at Blackwell and pass the up starting signal for Bromsgrove Station and the Lickey signal at Danger (if necessary).

A ticket must be issued to the Driver of the train engine and one to the Driver of the bank engine (if any). When a train is worked by two engines in front, the ticket must be shown to the Driver of the leading engine and given to the Driver of the train engine. If the train is assisted by more than one engine in the rear, the ticket for the bank engine Driver must be shown to the Driver of the leading bank engine(s) and given to the Driver of the rear bank engine.

If the apparatus fails after a train has left Bromsgrove but before it has passed the Lickey signal, Drivers must not proceed until verbal authority has been obtained by telephone from the Signalman at Blackwell.

If the telephone has failed, the authority must be obtained by the Fireman of the leading engine going forward to Blackwell box.

A Brakesman is on duty at Blackwell continuously except between 6am and 10pm on Sundays.

Down coal train photographed on 9th August 1950 descending the Lickey Incline behind L.M.S. 0-6-0 number 43891. The train would have paused at Blackwell to enable the brakes of the coal wagons to be 'pinned down' to arrest the progress of such a heavy consignment. The brakesman employed permanently at the summit had his work cut out to ensure all brake levers were 'pinned' correctly. The brakesman's post was dispensed with by B.R. in 1981. *(John Edgington)*

General Regulations

(i) Class 'A' and 'B' Passenger Trains, Class 'C' Express Freight Trains and Class 'C' Trains composed of Coaching Stock (not conveying passengers) must not exceed a speed of more than 10 miles per hour when passing through Blackwell Station, and must not descend the incline in less than five minutes.

(ii) Class 'D' Express Freight Trains and other Freight Trains below Class 'D' with not less than one third of the vehicles fitted with the continuous brake must STOP at Blackwell. The train may then proceed on the Brakesman's instructions without the application of hand-brakes, and must not descend the incline in less than seven minutes.

(iii) Class 'E', 'F', 'H', 'J' and 'K' Freight Trains not having one third of the vehicles fitted with the continuous brake must be brought to a stand at the top of the incline. If the loading is equal to less than eight wagons of mineral, the Driver must not proceed until he has received permission from the Brakesman to descend the incline. If the loading is equal to less than eight wagons of

Another job over! Two ex-L.M.S. 'Jinties' descend Lickey towards Bromsgrove on 7th July 1956 after banking yet another heavy train over Britain's steepest main line gradient. According to the rule (vi) on this page the engines are not coupled but are just buffered up. *(G. W. Morrison)*

mineral, the Driver must not proceed until he has received permission from the Brakesman, or the Guard of the train when the Brakesman is not on duty; the train may then descend with the application of hand-brakes. The train must not descend the incline in less than twelve minutes.

(iv) Class 'G' Light Engines and Engines with not more than two brake vans MUST BE BROUGHT TO A STAND at the top of the incline; they may then proceed without the Brakesman's permission.

(v) No train must exceed a speed of 10 miles per hour through Bromsgrove Station.

(vi) Except as shown in the following paragraph, when two or more banking engines are at Blackwell and require to return to Bromsgrove, they must be brought together at the top of the incline and the hand brake on the leading engine must be applied. The rear engine or engines must then propel the leading engine on to the incline and the brakes on all the engines must be used to prevent the engines being separated until they arrive at Bromsgrove.

During any failure of the down line block instruments, bells or track circuits, between Blackwell Station and Bromsgrove South signal boxes, banking engines must not descend the incline more than one at a time unless they are coupled.

(vii) The subsidiary arm, which is applicable only to bank engines, fixed below the Down Advanced Starting Signal for Blackwell Station will, when, taken off, display a letter 'C' or a letter 'W' illuminated during darkness. When the letter 'W' is displayed the signal is functioning as a warning signal (Rule 45) and when the letter 'C' is displayed the signal is functioning as a Call-on signal (Rule 44).

(viii) It must be distinctly understood that trains worked down the incline under Regulation (ii) above, must have the automatic vacuum brake operative on not less than one third of the wagons, all of which must be next to the engine, and the guard will be responsible for seeing that the requirements in this respect are complied with. Before the train leaves the last making-up station, he must satisfy himself or receive assurance from the person in charge that the vacuum brake is in working order and that at least 17 inches of vacuum are registered on the gauge applied to the last pipe of the fitted portion attached to the engine, otherwise the regulations in respect of the application of hand brakes before descending the incline must be strictly adhered to in accordance with the brake power instructions overleaf.

(ix) With trains on which the hand brakes have been applied, Drivers must start at such a speed as to enable this to be done.

(x) The Brakesman is responsible for instructing Drivers in accordance with paragraphs (ii) and (iii) above, before they descend the incline. When a brakesman is not on duty, Guards must satisfy themselves that the regulations have been complied with before giving Drivers the signal to start down the incline.

(xi) Down Line. The absolute block regulations apply (see paragraph vi) above, respecting the arrangements for returning bank engines.

Brake Power

(i) No train must leave Blackwell for Bromsgrove with less brake power than shown below:-

(a) All passenger trains must be worked with the continuous brake in accordance with the instructions in the General Appendix.

(b) If the continuous brake of a passenger train is out of order, the instructions in the following paragraph (c) must be complied with.

(c) Parcels, Fish, Fruit, Horse, Livestock, Meat, Milk, Pigeon or Perishable trains composed entirely of vehicles conforming to coaching stock requirements and empty coaching stock trains must have the number of brake vehicles shown below if the continuous brake is not in operation:-

	Number of braked vehicles if worked by one engine	**Number of braked vehicles if worked by two engines**
Not exceeding 7 vehicles	1	1
Exceeding 7 but not exceeding 11	2	1
Exceeding 11 but not exceeding 15	3	1
Exceeding 15 but not exceeding 19	4	1
*Exceeding 19 but not exceeding 23	5	2
*Exceeding 23 but not exceeding 25	6	3

*Trains exceeding 19 vehicles, but not exceeding 25 vehicles may descend the Incline with one brake van and one Guard, provided there are three engines in front.

(d) The braked vehicles must be brake vans in which a guard or Brakesman must ride, or vehicles fitted with hand brakes worked by side levers, or vehicles fitted with the continuous brake complete, in use from the engine.
NOTE If the number of braked vehicles on a train is not sufficient to comply with the above instructions, one or two bank engines, as may be necessary must be attached in front to control the train down the incline.

Heavy freight. Saltley-based 8F 2-8-0 number 48669 begins its decent of the Lickey Incline after pinning down of the brakes on an unidentied non-fitted freight at 4.30pm on 7th July 1956. *(G. W. Morrison)*

(e) The levers of all wagon brakes on freight trains must be dropped down and one wagon brake must be pinned down for every two loaded cattle, coal or heavy loaded goods wagons, and one for every three wagons or lightly loaded goods wagons.

(ii) The Brakesman must be careful to ascertain that the brakes pinned down are in good order, and he must increase the brake power beyond what is stated above, should he consider it necessary, in consequence of the state of the weather, or from any other cause. Brakesmen need not accompany trains down the incline, except when the number of guards is not sufficient to apply the prescribed number of hand brakes. Guards working freight and mineral trains must give what assistance is necessary to the Brakesman at Blackwell.

(iii) Guards of freight trains must apply the ordinary hand brakes when descending the incline, whether the vehicles in which they are travelling are fitted with a continuous brake in use from the engine or not.

General Notes on Descending.

(i) All down Freight trains except Classes 'C' and 'D' Freight trains and those with not less than one third of the vehicles fitted with the automatic brake must be brought to a stand between the passenger station and the South signal box with the rear vehicles clear of the connection between the down slow and the down fast line at the Bristol end of the station. In the event of a down Freight train coming to a stand before the rear vehicle is clear of the connection it must be drawn forward clear before the engine is uncoupled or any vehicle is attached or detached.

(ii) On arrival at Bromsgrove, any brakes that may have been applied must be released unless it is necessary for them to remain on for any other purpose.

(iii) No train, other than a Bank engine, must proceed on to the incline from Blackwell to Bromsgrove until the advanced starting signal has been taken off.

(iv) Calling On signals are fixed on the post of Bromsgrove Station down slow and fast starting signals, to give permission for Bank engines to proceed cautiously towards the South signal box when a train other than a passenger train is in the section, or to pass the starting signals at Danger for shunting purposes.

Down Freight Trains becoming divided in starting from Blackwell or in running between Blackwell and Bromsgrove.

(i) If a Freight train becomes divided in starting from Blackwell, arrangements must be made, after consultation with the Signalman and Brakesman, for the rear portion to be drawn back over the top of the incline to a point indicated by the Brakesman, and also for the front portion to be drawn or set back to the signal box where, if a vehicle is defective, it must be detached, the two portions re-coupled and the train allowed to proceed down the incline in the usual manner.

(ii) If it is not possible to re-couple the divided train, the rear portion must be left under the protection of the Brakesman, and the front portion carrying a tail lamp on the last vehicle and accompanied by the Guard must proceed down the incline provided a clear understanding has been made between the Guard, the Signalman and the Brakesman. The Guard must travel in accordance with the last Paragraph of Rule 183(d).

After the front portion has been cleared through the section and the line is again clear, the rear portion must be dealt with separately as a train and be worked down the incline with engine in front.

If no other guard is available to work the rear portion down the incline after the arrival of the first portion at Bromsgrove, the train Guard must return to Blackwell to accompany the second portion through the section.

(iii) If a Freight train becomes divided whilst descending the incline and the Guard has been successful in bringing the rear portion to a stand, arrangements must be made for the first portion to be disposed of at Bromsgrove. The Guard must secure the second portion of the train by pinning down all wagon brakes. The Driver must send his Fireman to the Guard to obtain Wrong Line order Form 'A' in accordance with Rule 183(f), the engine must then return in the wrong direction to the rear portion of the train and work to Bromsgrove in the usual manner.

If on returning to the second portion, the train engine cannot be coupled to the leading vehicle, arrangements must be made for an assistant engine from Blackwell to propel the second portion through to Bromsgrove with the train engine leading to control the movement and ensure all buffers maintain contact.

Hand brakes must be applied on sufficient vehicles to ensure the movement being made with safety.

The brakesman's entrance gate at Blackwell, with just the odd cautionary notice! — 2.6.89 *(John Ricketts)*

A period study of steam and telegraph poles, now both absent from the Lickey scene. 4F 0-6-0 number 43949 from Saltley shed pounds away, with assistance at the rear from a single 'Jinty' 0-6-0 (number 47638). The date is 13th April 1957 and it was Easter school holidays for the many young trainspotters who congregated at Bromsgrove, Vigo or Blackwell. *(Gerald Adams)*

An Observation ... Saturday 8th August 1959

This is the record of a two and a half hour period spent on the Lickey Incline at Vigo, on a Summer Saturday, during the height of the holiday season. The observer was Gerald Adams.

TIME	LOCOMOTIVES 'Up'	LOCOMOTIVES 'Down'	LOAD	TRAIN
2.38pm	44966 + 92231 + 8404		12 coaches	unidentified express
2.40pm		45040	10 coaches	unidentified express
2.43pm	44962 + 8400 + 5226		10 coaches	express to Leeds
2.54pm	73047 + 8402 + 8405		12 coaches	The Pines Express
2.56pm		45272 44839	12 coaches	unidentified express
3.07pm	73090 + 8401 + 8493		11 coaches	unidentified express
3.08pm		44804	5 suburban coaches	local passenger service
3.12pm	45682 + 8404 + 92231		12 coaches	The Devonian
3.14pm		44045	42 wagons (19 fitted)	freight
3.22pm	45608 + 5226		9 coaches	passenger to Sheffield
3.30pm	44853 + 8400 + 8405		11 coaches	passenger to Sheffield
3.38pm	44943 + 8403		8 coaches	passenger to Sheffield
3.47pm	92164 + 8401 + 8404		12 coaches	passenger to Leicester
3.56pm	44983 + 8402 + 5226		11 coaches	passenger to Derby
4.02pm		92008	8 coaches	unidentified express
4.04pm	73000 + 92231		12 coaches	passenger to Bradford
4.32pm	44666 + 8400 + 8405		11 coaches	passenger to Manchester
4.39pm	44167 + 8401 + 8404		9 coaches	passenger to Nottingham
4.48pm	73073 + 8402 + 5226		10 coaches	passenger to Sheffield
5.00pm	45660 + 92231		12 coaches	passenger to York
5.06pm		45685	10 coaches	passenger to Bristol
5.08pm	42818 + 8400 + 8405		10 coaches	passenger to Nottingham

Note: not recorded were the movements of the banking engines returning light from Blackwell to Bromsgrove. Class 9F 2-10-0 number 92231 had passed the observer four times in the up direction during this period, so had returned three times back down the Incline. Add to this engine the other active seven (8400-8405/5226) and their down journeys, coupled probably in twos, threes, or even fours, and one can see that linesiding the Lickey Incline on a summer Saturday was far from being dull!

Of interest is that more trains are travelling up, rather than down. This is due to the fact that most summer Saturday services went to the coast (i.e Bournemouth, Paignton, Newquay, etc) in the morning and returned later in the day to the industrial heartlands where the well-rested holiday makers lived, such as Sheffield, Manchester and Nottingham.

* *

3.12pm Jubilee 4-6-0 number 45682, 'Trafalgar,' banked by 8404 plus 92231, runs onto the top part of the Incline. The maroon liveried twelve coach load is mainly formed of B.R. Mark I coaching stock, with just an odd L.M.S. design in the centre. Train description boards adorn the roof-line of most carriages.

(Gerald Adams)

3.30pm. Another Stanier-designed 4-6-0 climbs past heading to Sheffield with a varied collection of stock, with L.M.S., L.N.E.R. and B.R. examples apparent. Black Five number 44853, a Leeds Holbeck engine, has two 0-6-0 pannier tanks assisting at the rear. One wonders if the British Transport Police ever walked the patch to give the young trainspotters a telling off for dangerous habits? *(Gerald Adams)*

3.47pm. It's all work at the backend as 9F 2-10-0 number 92164 passes by on the way to Leicester. All the carriages are in maroon colours other than the fourth L.M.S. type in 'blood and custard' livery. *(Gerald Adams)*

Freight Trains Booked to be Banked up the Lickey Incline

(for the period October 2nd 1989-May 13th 1990)

Times				Reporting Code	Train	Type
Bromsgrove		Blackwell				
Weekdays						
0003	0008	0017	MO	7E89	20.15(Sun) Briton Ferry — Immingham NCB Terminal	m.g.r
0129	0134	0143	MX	7E16	(22.15(FSX) Oakdale Colliery — Scunthorpe BSC (22.15(SO) Oakdale Colliery — Doncaster Decoy	m.g.r
0220	0225	0234	MX	6L88	20.00(SX) Trostre Works — Whitemoor New Yard	Speedlink
0335	0343	0352	MX	6E11	00.01(M-F) East Usk Jn New Yard — Belmont Yard (Doncaster)	Speedlink
1925	1933	1942	FSX	4M31	17.20(FSX) Swindon Cocklebury — Longbridge East	Company
2112	2120	2129	SX	6E64	18.03(M-F) East Usk Jn New Yard — Tinsley New Yard	Speedlink
2158	2203	2212	SX	7E95	18.30(M-F) Oakdale Colliery — Scunthorpe BSC	m.g.r
2220	2227	2236	MWFO	6E45	19.25(MWFO) Severnside ICI — Haverton Hill	Company
2320	2325	2334	MSX	6E44	20.40(MSX) Cardiff Tidal Sidings — Scunthorpe West	Company
Sundays						
2159	2206	2215	SUN	7E87	17.35(Sun) Briton Ferry — Immingham NCB Terminal	m.g.r

The freight trains quoted above were diagrammed for banking at the commencement of the winter 1989/90 timetable, but due to the rapid, ever-changing requirements of today's freight business on British Rail it may be the case that not all trains run. In fact, additional trains not listed may have run!

The banking requirements for this period call for two class 37 locomotives in multiple, provided from Gloucester stabling depot, and manned by drivers from there.

During this Winter timetable period no passenger or parcel trains were diagrammed for banking assistance.

Two Class 37/0s, numbers 37165 and 37210, are seen awaiting their next banking duty at Bromsgrove on the 8th August 1985. Nowadays, all requirements for banking are during the evening and night time. *(C. J. Tuffs)*

GOOD NEWS FOR BROMSGROVE

Starting 15 May 1989 extra services will be operating on the Worcester - Droitwich - Bromsgrove - Barnt Green line. Passengers travelling from Worcester, Droitwich or Bromsgrove will be able to reach University, Five Ways, Birmingham New Street etc. more frequently by changing at Barnt Green. There will be a new direct service between Worcester and Bromsgrove (fastest journey time 18 minutes) and if it's shopping in Redditch that takes your fancy - that's easier to!

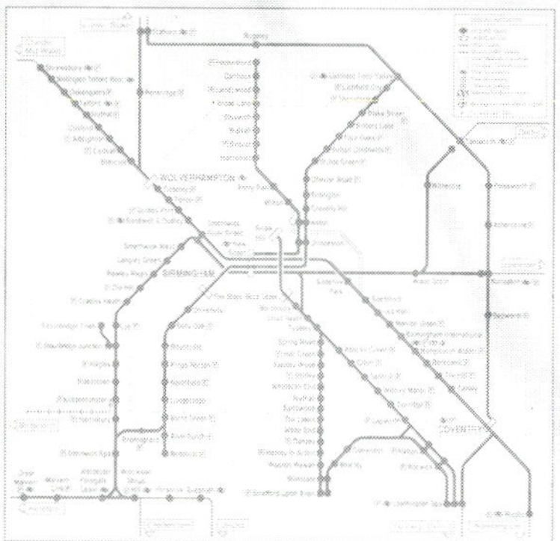

MIDLINE
SIMPLY THE BETTER WAY

GOOD NEWS FOR BROMSGROVE

FARES FROM BROMSGROVE TO

	Single	Cheap Day	Off Peak*
Worcester	1.40	1.90	1.50
Droitwich	1.00	1.50	-
Redditch	1.80	2.30	1.60
Birmingham	2.00	2.60	1.70

FARES TO REDDITCH FROM

Worcester	3.20	4.40	2.40
Droitwich	2.40	2.80	2.20
Bromsgrove	1.90	2.30	1.60

DAY RANGER FARES	Adult	Child	Family
	4.50	2.25	9.00 2 adults & up to 4 children

A Day Ranger ticket will give you the freedom of the MIDLINE network. For further details enquire at your local staffed station or ring 021-643 2711.

*Off peak fares apply Mon to Fri after 0845 all day Saturday and Bank Holidays.

MIDLINE
SIMPLY THE BETTER WAY

Publicity leaflet produced for the improved MIDLINE passenger services serving Bromsgrove, effective from 15th May 1989.

The Monday to Saturday timetable for the 2nd October 1989-13th May 1990 timetable period, with eleven services in each direction. There are no Sunday services.

Barnt Green–Bromsgrove–Worcester

Mondays to Saturdays

	A				SX	SO	SX	SX	SO		SO
Birmingham New Street d	0706	0936	1106	1306	1506	1506	1721	1751	1736	1954	2206
Five Ways d	0709	0939	1109	1309	1509	1509	1724	1754	1739	1909	2209
University d	0713	0943	1113	1313	1513	1513	1728	1758	1743	1913	2213
Redditch d	0753	1023	1153	1323	1553	1553	1653	1753	1753	1953	2223
Barnt Green d	0811	1041	1211	1349	1611	1611	1751	1823	1823	2011	2243
Bromsgrove d	0816	1046	1216	1354	1616	1616	1756	1828	1828	2016	2246
Droitwich Spa d	0826	1056	1229	1404	1626	1626	1806	1838	1838	2029	2256
Worcester Shrub Hill a	—	—	1237	1412	—	1634	1814	—	—	2037	2304
Worcester Foregate Street a	0835	1105	1241	1416	1635	—	—	1847	1847	—	—

A-To Great Malvern arr. 0847
SO-Saturdays only
SX-Saturdays excepted

Worcester–Bromsgrove–Barnt Green

Mondays to Saturdays

	SX	SX	SX	SO							SO
Worcester Foregate Street d	—	—	—	—	0948	1131	1300	1500	1732	1900	2202
Worcester Shrub Hill d	—	—	0728	0728	0953	—	—	—	—	—	—
Droitwich Spa d	—	—	0736	0736	1005	1140	1309	1509	1741	1909	2211
Bromsgrove d	0647	0715	0751	0751	1015	1150	1322	1518	1750	1919	2220
Barnt Green a	0655	0723	0800	0800	1023	1158	1330	1526	1758	1929	2228
Redditch a	0718	0748	0818	0818	1048	1218	1348	1548	1818	1948	2248
University a	0728	0758	0821	0828	1058	1228	1358	1558	1828	1958	2258
Five Ways a	0732	0802	0825	0832	1102	1232	1402	1602	1832	2002	2302
Birmingham New Street a	0736	0806	0829	0836	1106	1236	1406	1606	1836	1948	2306

SO-Saturdays only
SX-Saturdays excepted

Bibliography

The Railway History of Bromsgrove and the Lickey Incline:		
Part I 1840-1948	R. Harris	
Part II 1949-1982	R. Danes	
The Lickey Incline	H. C. Casserley.	Oakwood Press
The Midland Railway	C. H. Ellis.	Ian Allan
The Bristol and Gloucester Railway	G. C. Maggs.	Oakwood Press
The Birmingham-Gloucester Line	G. C. Maggs.	Line One
The Birmingham and Gloucester Railway	P. J. Long/Rev. W. V. Awdry.	Alan Sutton
The Railway Magazine (October 1904)		

A Photographic Ending

Blackwell

The part brick, part wooden constructed signal box at the Birmingham end of Blackwell Station is the common feature in these two photographs.

(Upper) Stanier 'Jubilee' class 4-6-0 number 45725 'Repulse' accelerates away from the Lickey summit with a West of England to the north express in the late 1950s. *(B. S. Moone)*

(Lower) A few years later, in 1964, a newly delivered Sulzer type 2 Bo-Bo diesel electric number D7589 drifts into the down platform with a Birmingham New Street-Worcester tea-time stopping service. *(Tony Haynes)*

Just Visiting

(Upper) B.R. Standard Pacific 4-6-2 number 72005 'Clan Macgregor' storms through Blackwell on the 07.45am Paignton-Newcastle train, working via Worcester Shrub Hill, on 9th July 1960. The shedplate is 12A – Carlisle Upperby. Why was a Carlisle allocated locomotive working so far south of its home shed, one wonders?

(W. H. A. Thompson)

(Centre) On 24th April 1985 a class 151 multiple unit ran two test runs to simulate passenger diagrams between Lichfield City and Bromsgrove, which included a standing start on each Bromsgrove to Lichfield trip on the Lickey Bank. On the second trip 151.001 ran with one engine isolated and is pictured here making a standing start at Milepost 54¼. *(G. Bannister)*

(Lower) Class 40s numbers 40057 and 40084 are 'spotted' ascending the Lickey Incline at Vigo on the 8th May 1982 in charge of the Severn Valley Railtours 'Cotswold Venturer'. They are seen on the return journey from London Paddington to York.
This class of locomotive was never a regular performer over the Lickey and by the way each available coach window is occupied the passengers are very receptive to the noise the hard-working 'whistling' engines are making!

(C. J. Tuffs)

Passenger Power

Three decades of passenger train working are illustrated on this page, and reinforce the fact that 'human' travel has always been the main source of train movements over the Lickey.

(Upper) Coasting down 'wrong line', due to engineering work in the vicinity, is green liveried 'Peak' type 4 number D162, with the southbound 'Cornishman' on the 30th August 1964. Eleven of the B.R. Mk1 coaches are in the then common maroon livery, whilst the odd-man-out is a Western Region chocolate and cream liveried variation. *(Michael Mensing)*

(Centre) Not long after 9am on the 21st May 1979 and class 47 number 47091 is on full power as its Sulzer engine strains with nine coaches behind. The service is the 07.30 Bristol Temple Meads to Manchester Piccadilly. *(Kevin Connolly)*

(Lower) This study typifies current British Rail operations over the Lickey. A High Speed Train set, forming the 07:00 Newcastle to Plymouth InterCity service, zooms southwest on the 8th August 1985. The well-known Blackwell Sanitorium forms a familiar backdrop. *(C. J. Tuffs)*

(Overleaf) A classic of man and machine. *(Gerald Adams)*

Man and machine, in harmony—
OVER THE LICKEY